CRUISING THE GEOLOGIC TIME IN THE SAN JUAN ISLANDS

The Geology of the San Juan Islands

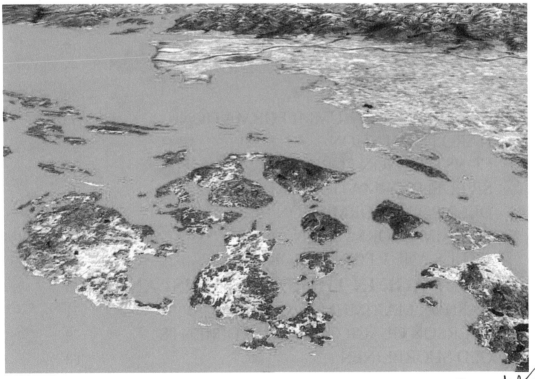

Don J. Easterbrook

To Kathy & Roger
with best wishes
Don

TABLE OF CONTENTS

INTRODUCTION

I wasn't born in the San Juan Islands, but I got there as soon as I could. I first became acquainted with the San Juan Islands as a high–school student and became enchanted with them. After completing my undergraduate degree at the University of Washington, I returned to the islands in 1958 for a summer of field study as part of my Master's degree program. In the 1960s, I received two National Science Foundation research grants and spent several field seasons studying the geology of the islands, leading to publication of several papers, mostly dealing with the Ice Age history of the region. In the following decades, I visited the islands from time to time and published several more papers on the Ice Age in the San Juan Islands. My interest in the islands was rekindled in recent years as a result of new technological advances in laser, sonar, and satellite imagery. Laser beams from airplanes can now be bounced off the ground surface and computers can convert the recorded data into images resembling air photographs. The laser beams penetrate vegetation on the ground and the result is an image of the Earth's surface without any vegetation. The ground surface is literally laid bare, and this is hugely valuable to geologists because geologic features that are obscured by vegetation become easily visible. This new technological breakthrough (called LIDAR) has given geologists an incredibly valuable new tool for understanding geologic features. In 2012, I obtained LIDAR imagery, sonar imagery of the sea floor, and satellite imagery of the San Juan Islands and began a re-evaluation of the geology of the islands. From the beginning of the project, LIDAR, sonar, and satellite images were stunning, revealing many geologic features that had never been observed before because of vegetation. In particular, the LIDAR and sonar imagery exposed numerous high–angle faults in locations that were previously believed to be low–angle, thrust faults. As a result, it is now apparent that the entire geologic framework of the San Juan Islands, unchallenged for 40 years, is in fact, not valid and requires major revision.

In the course of this project, I reviewed everything published in professional journals on the geology of the San Juan Islands. What became apparent was that a lot of good geology has been published in many short journal articles, but there was no single compilation of the geology of islands as a whole, and virtually nothing was available for non-geologists (island residents, boaters, tourists). So in 2012, I set out to compile as much geologic information as possible and put it into a single book for both geologists and non-geologists. Portraying the geology of the San Juan Islands for both geologists and non-geologists turned out to be more difficult than it might seem—it's like trying to write in two different languages at the same time. Geologists have technical terms for everything, (most of which are likely incomprehensible to non-geologists), so trying to translate all technical terms and concepts into basic English is a daunting task and some kind of compromise is called for. The first part of the book is intended for non-geologists and every effort was made to use plain English. The middle part of the book contains the new LIDAR, sonar, and satellite data published for the first time and is intended primarily for geologists (although, hopefully, it is also understandable to non–geologists). Non-geologists may want to skip over parts of it and move on to the last part of the book, which summarizes the geology of all of islands individually with many photographs, detailed geologic maps, and LIDAR, sonar, and satellite images.

Organization of chapters

The geologic features of the San Juan Islands described in the book are arranged in several sections: (1) introduction and basic concepts, (2) the geology of rocks in the islands arranged from oldest to youngest; (3) new LIDAR, sonar, and satellite data and its implications, (4) geology of the individual islands; (4) an appendix of sonar imagery of the sea floor, (5) glossary of geologic terms; (6) references, and (7) an index. An attempt has been made to focus on geologic features that can be seen in the field. Descriptions of specific localities where interesting geologic features can be seen are included throughout the book.

THE SAN JUAN ISLAND ARCHIPELAGO

The San Juan Islands consist of an archipelago in the Salish Sea of the northern Puget Lowland in Washington (Fig. 1). At Friday Harbor, the tide ranges from lows of –2 to highs of +9.5 feet, so the total number of islands and reefs above sea level varies greatly every day. At the lowest tides, nearly 800 islands and reefs are above water, but at high tides the number is only about half that. About 170 named islands are permanently above sea level.

The San Juan Islands are not only scenic, but are also among the most interesting geologically. A wide variety of rocks, some dating back half a billion years, hold a fascinating history of major geologic changes. Recorded in the rocks of the region is an enthralling tale of major crustal upheavals, intrusion of molten rock, and burial by thousands of feet of glacial ice.

The San Juan Islands are the highest points of submerged peaks that extend from the mainland to Vancouver Island. The larger islands are rugged and mountainous, with fifteen peaks having elevations exceeding 1000 feet, typically rising precipitously from the water's edge. Channels between islands attain depths of 600 feet and in places exceed 1000 feet. The highest point is Mt. Constitution with an altitude of 2,409 feet. The deepest water in the San Juan Islands is 1,356 feet below sea-level in Haro Strait near Stuart Island, giving a relief of 3,765 feet.

Figure 1. The San Juan Islands

For purposes of definition in this book, all of the islands in the Salish Sea north of the Strait of Juan de Fuca, east of Haro Strait, south of Georgia Strait, and west of the mainland, are considered part of the San Juan Islands. Many other definitions are possible, but these are the islands that most residents, boaters, and visitors consider the San Juan Islands.

DISCOVERY AND NAMING OF THE SAN JUAN ISLANDS

In 1592, Juan de Fuca, a Greek sea captain who sailed the Pacific in the late 1500's for the Spanish, discovered a "broad inlet of sea" along the coast of the Pacific Northwest. In 1778, British explorer Captain James Cook sighted Cape Flattery at the entrance to the Strait of Juan de Fuca, but didn't enter the strait. In 1787, maritime fur trader Captain Charles W. Barkley explored the entrance to the strait and named it the Strait of Juan de Fuca.

In 1790, Spaniard Manuel Quimper charted and named many geographic features along the northern and southern shore of the strait and saw the San Juan Islands but mistook them for part of the mainland to the north. He reported seeing Mt. Baker, which he named La Gran Montagna Carmello. Quimper stopped in Bellingham Bay near the mouth of Padden Creek in 1791. Most of his discoveries along the strait were later renamed by George Vancouver. Among those which retained Spanish names include Port Angeles, Rosario Strait, Quimper Peninsula, and Fidalgo Island.

5

The name "San Juan Islands" originates from the 1791 expedition of Francisco de Eliza, who claimed the islands for Spain and named them Isla y Archiepelago de San Juan. San Juan Island was discovered by Gonzalo López de Haro, one of Eliza's officers. Eliza named Haro Strait and Orcas Island for him.

In 1792, British Captain Vancouver explored the inland waterways and kept many of the Spanish names, but renamed others. In 1841, American explorer Charles Wilkes gave new names to a large number of coastal features. Because of the confusion of multiple names, the British Admiralty unified names on official British charts, including the San Juan Islands. Most of the British and Spanish names were kept, but most of Wilkes' names were replaced except for some named after American naval officers, such as, Shaw, Decatur, Jones, Blakely, and Sinclair.

PREVIOUS WORK

The earliest geologic observations in the San Juan Islands date back to 1857, focusing on fossils in sandstone on Sucia, Spieden, and Stuart Islands. The first comprehensive study of the geology of the islands was the PhD thesis of Roy McClellan at the University of Washington (UW). He began field work in the islands in 1924 and spent several months over the next three years rowing a boat from island to island to map the geology. His thesis was published in 1927 (Fig. 2).

For more than 30 years, this remained the only source of geologic information on the San Juan Islands until 1957 when Ted Danner completed a PhD thesis at the UW on fossils and geologic relationships of rocks in the islands. He continued working on fossils and the rocks that contained them for many years thereafter.

My studies of the geology of the islands began in 1958 when I spent the summer with a group of geology graduate students and faculty from the University of Washington and mapped the eastern part of Orcas Island as part of my Master of Science program. My geologic map and report on *"The geology of eastern Orcas Island and Sucia Island"* (Fig. 2) made significant revisions to McLellan's work, the most significant of which was recognition that the Turtleback Crystalline Complex was a basement complex of ancient rocks, not igneous intrusions as mapped by McLellan. The report also revised much of the stratigraphy and named a number of new rock units.

Another 17 years passed before new geologic maps and reports of the San Juan Islands were published by Joe Vance and John Whetten of the UW in 1975. In the years that followed, published papers by Vance, Whetten, Brandon, Cowan, Johnson, Brown, Blake, and Engebretson plus many MS theses from the University of Washington, Washington State University, and Western Washington University have added to our understanding of the geology of the islands.

Since 2012, I've been using new technological advances in laser, sonar, and satellite imagery to study the geology of the San Juan Islands. To my great surprise, these new techniques have shown that interpretation of the primary geologic structure of the islands as five extensive thrust faults, unchallenged for 40 years, is in fact, not valid. The laser and sonar imagery discussed later in the book demonstrate that the five thrust faults that make up the San Juan Island Thrust System, the Rosario, Orcas, Haro, Lopez, and Buck Bay thrust faults, have been incorrectly mapped and do not actually exist.

ACKNOWLEDGEMENTS

My earliest experiences with the geology of the San Juan Islands were enhanced by Joe Vance, Ted Danner, John Whetten, and Ross Ellis. John Whetten kindly reviewed an early draft of the manuscript and offered many helpful suggestions. George Mustoe generously contributed a number of excellent photographs.

Thanks to Karen Easterbrook Sutton and Alex Sutton for many hours of exploring the San Juan Islands aboard their boat and to Alex for many hundreds of photographs.

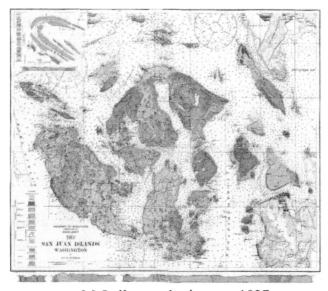

McLellan geologic map, 1927

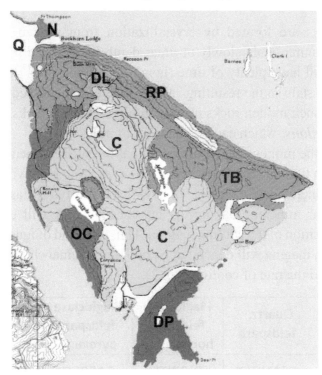

Easterbrook geologic map of eastern Orcas Island, 1958.
TB=Turtleback Complex, RP=Raccoon Pt. Fm., DL=Day Lake Fm.,
OC=Orcas Chert, C=Constitution Fm., DP=Deer Pt. Fm., N=Nanaimo Fm.,
Q=Quaternary

Figure 2. Two of the earliest geologic maps in the San Juan Islands.

SOME BASIC GEOLOGIC CONCEPTS

The geologic history of the San Juan Islands is written in its rocks and landforms. All we need to do is read it. To do that requires some knowledge of basic geologic concepts. This chapter is a very brief summary of some of the concepts that will help reveal the geologic history of the islands.

ROCKS AND MINERALS

Minerals are inorganic, naturally occurring, crystalline solids. Each mineral has a particular internal crystal structure which is always the same for that particular mineral. Some examples of common minerals are quartz, feldspar, biotite mica, hornblende, pyroxene, olivine, and calcite. Rocks are composed of two or more minerals. Three kinds of rock occur in nature: igneous, sedimentary, and metamorphic.

IGNEOUS ROCKS

Igneous rocks are formed by crystallization from molten rock, known as magma. If a magma cools slowly at great depth below the Earth's surface, the resulting rock will have plenty of time (hundreds to thousands of years) in which to grow large crystals so the resulting rock will be coarse grained (crystals ¼ inch or so). Because such molten rocks intrude into pre-existing rocks, they are known as *igneous intrusions*, which can be hundreds of miles across.

However, if the molten rock cools rapidly at the surface, then the crystals will not have as much time to grow very large, so the resulting rock will be fine grained or glassy, and minerals will be visible only with a magnifying glass or microscope.

The particular minerals making up any igneous rock will depend upon the chemical composition of the cooling magma. Thus, the kind of igneous rock formed from cooling of a magma will depend upon both the original chemical composition of the magma and the rate of cooling (Fig.2).

Minerals¤	Quartz,· feldspar¤	Plagioclase· feldspar,· hornblende¤	Plagioclase· feldspar,¶ pyroxene¤	Pyroxene,· olivine¤	Olivine¤
Coarse·¶ grained¤	GRANITE¤	DIORITE¤	GABBRO¤	PERIDOTITE¤	DUNITE¤
Fine·¶ grained¤	RHYOLITE¤	ANDESITE¤	BASALT¤	¤	¤
Color¤	White	Gray	Black	Green	Green

Figure 2. Types of igneous rocks.

Granite (Fig. 3) is composed of the minerals quartz and feldspar with smaller amounts of biotite or hornblende.

Diorite (Fig. 3) is similar to granite except that it does not contain quartz.

Gabbro is composed of plagioclase feldspar and pyroxene with some olivine. It is usually darker than granite.

Peridotite is composed of olivine and pyroxene and is usually green colored.

Dunite is composed entirely of olivine and is also usually green.

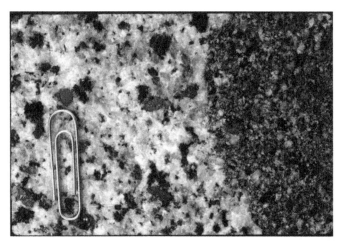

Figure 3. Granite (left) and diorite (right). The light colored minerals in the granite are quartz and feldspar and the dark mineral is biotite. The diorite contains feldspar, but no quartz and is darker because it contains black hornblende.

Fragmental material blown out of volcanoes is known as **tuff** if the particles are sand size or less and **breccia** if the particles are pebble size.

The most common lava in the San Juan Islands is **basalt**, which is the surface equivalent of gabbro. When basalt forms under water, it cools quickly, forming ellipsoidal masses about the size of pillows. Such lavas are known as **pillow basalts**.

SEDIMENTARY ROCKS

Sedimentary rocks are formed at the Earth's surface by deposition of sediment and precipitation of chemicals in the ocean. Gravel deposits form **conglomerate** when the pebbles become cemented together. Sand which later becomes cemented forms **sandstone**. Compacted silt and clay forms **argillite (shale)**.

Limestone forms by precipitation of chemicals in sea water or by the accumulation of organisms such as clams, corals, and other marine organisms. Chemical precipitation of micro crystalline quartz forms **chert**, which can also form by the deposition of radiolarians, microscopic organisms whose shells are made of silica.

Grain size		Sedimentary Rock	
	Gravel	Conglomerate, breccia	
	Sand	Sandstone	
	Silt	Argillite (shale)	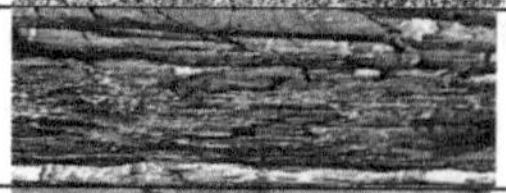
	Clay		
Chemical precipitation of calcite		Limestone	
Accumulation of fossils with calcite shells			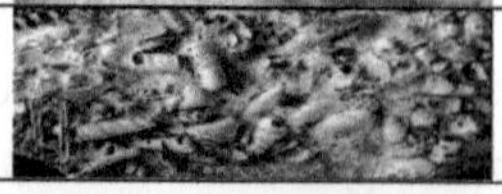
Chemical precipitation of silica		Chert	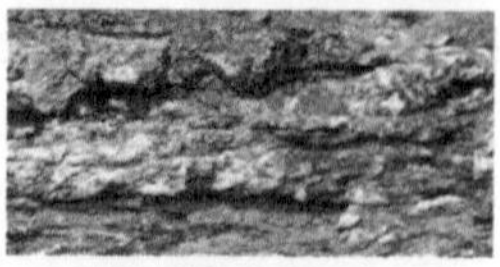
Accumulation of micro-fossils with silica shells			
Accumulation of plant remains in peat bogs		Coal	

Figure 4. Types of sedimentary rocks.

The mineral composition of sandstone is determined by the rocks being eroded in the source area and by depositional processes. Two types of sandstone occur in the San Juan Islands, **greywacke** and **arkose**.

Greywacke sandstone (Fig. 5) is composed of poorly sorted, angular grains derived from erosion of volcanic rocks, indicating relatively short water transport. It is abundant in the islands with thicknesses of thousands of feet

Arkosic sandstone (Fig. 5) consists mostly of grains of quartz and feldspar which are more rounded and better sorted than the grains in greywacke. Because sedimentary rocks form on the Earth's surface, they often show features that are invaluable in interpreting ancient environments and tectonic events.

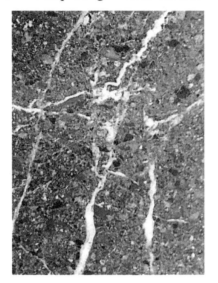 

Figure 5. **Greywacke sandstone** (left), is composed of volcanic, sand–size grains deposited as turbidites. Dark grains are volcanic fragments. Almost all of the Paleozoic to mid–Mesozoic sandstones are greywacke. **Arkosic sandstone** (right), is composed of feldspar and quartz sand–size grains derived from weathering and erosion of granitic source rocks. The Nanaimo and Chuckanut sandstones consist of this kind of sandstone.

Graded bedding (Fig. 6), is the progressive decrease in grain size toward the top of an individual bed, suggesting rapid deposition from turbid water in which the coarser grains settle out first.

Cross-bedding (Fig. 7), is a type of bedding in which the layers are inclined at an angle as a result of having been deposited on the backside of a sand or gravel bar.

Fossils. The remains of living organisms buried in sediment that are preserved as fossils.

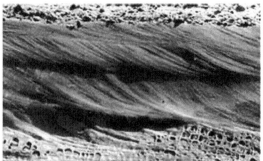

Figure 6. Graded bedding. Lower, coarse grains become finer upward.

Figure 7. Cross–bedded sandstone, deposited on sand bars.

METAMORPHIC ROCKS

Metamorphic rocks are formed by recrystallization of preexisting rocks as a result of increased heat and pressure. The nature of a metamorphic rock depends on the composition of the original rock and the amount of heat and pressure applied to it. The heat and pressure produces new minerals, new textures, and new structures in rocks.

Shale subjected to modest heat and shearing pressure will be transformed into a **phyllite**. With more intense heat and shearing pressure a **schist** is formed. If the temperature and pressure are even higher a coarse grained **gneiss** having a composition similar to granite, is formed. Metamorphism of basalt under low grade heat and pressure forms **greenschist**. Higher temperature may result in crystallization of plagioclase feldspar and hornblende to form **amphibolite**.

Original rock	Increasing temperature and pressure, recrystallization→			
SHALE	SLATE	PHYLLITE	SCHIST	GNEISS
BASALT	GREENSTONE	GREENSCHIST		AMPHIBOLITE
LIMESTONE	MARBLE			
SANDSTONE	QUARTZITE			

Figure 8. Types of metamorphic rocks.

Phyllite forms by recrystallization of shale by low grade increase in heat and pressure to form small crystals of quartz and mica and develop foliation from shearing forces.

Schist forms by further recrystallization to form quartz, feldspar, mica, and garnet with higher temperatures and more strongly developed foliation from shearing.

13

Gneiss forms by further recrystallization to form quartz and feldspar (similar to granite) with mica, garnet, hornblende, and pyroxene accessory minerals and strongly developed coarse banding foliation from shearing.

Greenstone forms from recrystallization of basalt to convert pyroxene and feldspar into green minerals such as actinolite, chlorite, epidote, serpentine, and talc by low grade increase in heat and pressure.

Greenschist forms from recrystallization of basalt by low grade increase in heat with strong shearing pressure that results in stronger foliation. Its green color comes from recrystallized actinolite, chlorite, epidote, serpentine, and talc.

Ampibolite forms from still higher heat and pressure grade to form feldspar, hornblende, and pyroxene.

FOLDING

Crustal forces may deform rocks into **folds**. Horizontal bedding may become folded into arches (anticlines) and troughs (synclines). The orientation in space of a tilted bed can be described by the compass direction of a horizontal line on a bedding plane (strike or trend) and the angle of inclination of the bedding (Fig. 9).

Anticlines (Fig. 10). Beds dip away from the central axis of the fold.

Synclines (Fig. 10). Beds dip toward the central axis of the fold

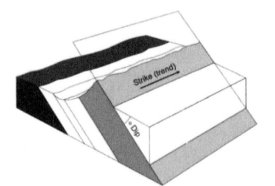 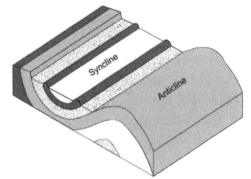

Figure 9. Dip and strike (trend) of a tilted bed.

Figure 10. Anticline and syncline. Beds in the anticline dip away from the central axis. Beds in the syncline dip toward the central axis.

FAULTING

Faults are fractures in rocks along which movement has occurred on opposite sides of the fault. Two types of faults occur in the San Juan Islands–high-angle faults, and thrust faults.

Normal faults are high–angle faults in which movement of the upper block is down the dip of the fault plane (Fig. 11).

Thrust faults are low angle faults which push the upper plate over the underlying rocks (Fig. 12).

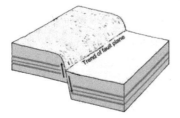 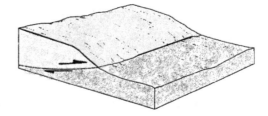

Figure 11. High-angle normal fault. Movement has occurred down the dip of the fault plane.

Figure 12. Low angle thrust fault. The upper plate has been pushed over the underlying rocks.

GEOLOGIC TIME

The length of geologic time is immense—about 4 ½ billion years. Since the Earth formed, it has undergone countless changes, many of which are still going on. For about the first 3–4 billion years, early Pre-Cambrian time, no life existed on Earth. Primitive forms of life appear in rocks for the first time about a billion or so years ago, and more complex forms appear in the fossil record beginning about 540 million years ago.

The earliest geologic time scales subdivided geology time into Eras and Periods based on the law of superposition (younger beds lie on older beds) and the law of faunal succession (each period of time is characterized by specific life forms that are unique to that time). The **Paleozoic Era** (paleo=early, zoic=life) was characterized by the dominance of invertebrate fossils with no backbone. The **Mesozoic Era** (meso=intermediate, zoic=life) was characterized by the dominance of reptiles, including dinosaurs. The **Cenozoic Era** (ceno=recent, zoic=life) was characterized by the dominance of mammals. Humans appear in the fossil record at the very end of the Cenozoic Era. Each Era has been subdivided into Periods.

Rocks from various time intervals in the Earth's history could be placed in the early geologic time scales on the basis of fossils and stratigraphy, but this was only **relative time** and no one knew the **numerical time** measured in years. Today, geologic time in years can be measured using radioactive isotopes, sometimes referred to as isotope clocks. **Isotopes** are atoms with the same number of protons in their nuclei as other species of an element but with a different number of neutrons in their nuclei. The nuclei of some isotopes are unstable and disintegrate spontaneously, giving off radiation in the process. If the number of protons in their nuclei changes, these parent elements become transformed into daughter elements at constant rates independent of temperature, pressure, or chemical processes.

Because of their constant rate of spontaneous disintegration, radioactive isotopes allow determination of age if the ratio of parent atoms to daughter atoms can be determined and if the rate of transformation is known. The three radioactive isotopes most useful for dating rocks in the San Juan Islands are uranium→lead (U-Pb), potassiuim–40→argon (^{40}K–Ar), and carbon–14→nitrogen. Many age determinations have been made of rocks and sediments in the San Juan Islands.

ERA	AGE (PERIOD)	MILLIONS OF YRS	GEOLOGIC UNIT		DESCRIPTION
CENOZOIC	QUATERNARY	2	Ice Age deposits		Glaciomarine drift
	TERTIARY	65		Chuckanut Fm	Arkosic ss
MESOZOIC	CRETACEOUS	150		Nanaimo Fm	Arkosic ss
			Deer Pt. Fm	Spieden Fm	Greywacke congl., ss
	JURASSIC	200	Constitution Fm	Carter Pt Fm	
			Orcas Chert		Ribbon chert
	TRIASSIC	250		Haro Fm	Greywacke ss
PALEOZOIC	PERMIAN	300	East Sound Group		Volcanic rocks, greywacke ss
	PENNSYLVANIAN	325			
	MISSISSIPPIAN	360			Limestone lenses
	DEVONIAN	420			
	SILURIAN	445			
	ORDOVICIAN	485	Turtleback Crystalline Complex		Amphibolite, granite, diorite, gabbro
	CAMBRIAN	540			
PRE-CAMBRIAN					

Figure 13. Geologic time scale.

ROCKS OF THE SAN JUAN ISLANDS

The San Juan Islands are composed chiefly of Paleozoic and Mesozoic sedimentary rocks, the oldest of which have been recrystallized by heat and pressure. The Paleozoic beds, which include the bulk of the sedimentary rocks of the central islands (Orcas, San Juan, Lopez, Shaw, Blakely, Decatur), have been intensely deformed. The eastern part of the map-area (Cypress, Guemes, Sinclair Lummi, Eliza) is composed of large irregular masses of peridotite. The islands forming the northern and northwestern margins of the map area (Stuart, Spieden, Waldron, Sucia, Matia, Patos) are composed of folded and faulted sedimentary rocks of late Mesozoic and early Tertiary age.

The islands were completely overridden by glaciers during the Ice Age, and even the top of Mt. Constitution is deeply grooved and polished by glacial scouring. Considerable areas on many of the islands are covered with glacial deposits. During the Ice Age, the islands were depressed by the weight of thousands of feet of ice. Removal of the weight of the Cordilleran Ice sheet when it melted at the end of the Ice Age resulted in uplift of the islands, and excellent examples of raised marine shorelines are to be seen in many places at elevations up to 300 feet.

TURTLEBACK CRYSTALLINE BASEMENT ROCKS— AN ANCIENT PIECE OF THE EARTH'S CRUST

The Turtleback Range of Orcas Island is so named because the rounded rock knobs that make up the range resemble the shape of a turtle (Fig. 14), with the main part of the range making up the body and Orcas Knob making the head (Fig. 15).

Figure 14. Turtleback Mt., Orcas Island

Figure 15. Orcas Knob, the 'head' of Turtleback Mtn.

The Turtleback crystalline complex was originally mapped by McLellan in 1927 as an igneous intrusion crystalized from molten rock as granite, diorite, and gabbro during the Jurassic Period (~200 million years ago). He also included peridotite and dunite of the Fidalgo Formation. No further geologic studies of the islands were made for 31 years, so this was the prevailing view when I began mapping the geology of eastern Orcas Island in 1958. In the course of field work along the eastern shoreline of East Sound, UW professor Joe Vance, another graduate student, and I discovered a rock outcrop of amphibolite in the Turtleback crystalline complex along the east side of East Sound that was clearly metamorphic, not igneous as mapped by McLellan. We mapped Turtleback gneiss, amphibolite, diorite, gabbro and other crystalline rocks along the shore of East Sound, along the coastal area NW of Point Lawrence, Buck Mt., Pickett Mt., and Mountain Lake east of Mt. Constitution. Amphibolite is the most abundant rock type, but diorite is also common. Turtleback gneiss is distinctly foliated (banded) and in places is intricately mixed with diorite and other fine grained crystalline rocks.

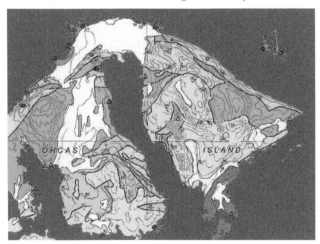

Figure 16. Extent of the Turtleback Complex (pink) on Orcas Island.

The first geologic description of the Turtleback as a pre–Devonian basement complex is shown below in italicized text (Easterbrook, 1958).

"Basement Complex

General features: Along the shores of East Sound, Point Lawrence, and Mountain Lake, diorite and amphibolite associated with gneiss and migmatites are seen. Previously, true metamorphic basement rocks were unknown on the island and all crystalline rocks had been mapped as intrusive. However, gneiss, amphibolite, and diorite occur on the shore of East Sound overlain by the Chert formation of Devono-Mississippi age. These rocks have been mapped as a basement complex of Pre-Devonian age which seems to outcrop continuously from Point Lawrence northwestward and thence southwestward to East Sound. They are overlain at East Sound by the Chert formation, on Mt. Pickett and Mt. Constitution by the Constitution formation, and on Buck Mt. by the Day Lake formation.

Lithology: The basement complex consists of several different rock types. Perhaps the most common rock in the basement complex are amphibolites, which are medium grained, gray-green in color, with crystals averaging from one to four millimeters. Plagioclase and quartz are present, but most of the rocks are dominated by amphibolite. Diorites area also common. They are medium grained, with about 60-70% plagioclase, 10-20% hornblende with some biotite and a little quartz. The migmatites consist of intricately mixed and scrambled medium grained diorite and finer grained crystalline rocks. The diorites occur in pods and chaotic layers and lenses within a finer grained crystalline matrix. In some places the diorite is dominant. The gneisses are distinctly banded or foliated in outcrop and are usually fine grained in a hand specimen. Some porphyries are found with plagioclase crystals up to a centimeter in diameter enclosed in a fine grained matrix.

Age: The basement complex is overlain by the Chert formation of presumable Devono-Mississippian age and is therefore Pre-Devonian. It may actually be anywhere from Pre-Cambrian to Devonian in age."

The discovery that the Turtleback crystalline complex contained pre-Devonian metamorphic rocks required major changes in the interpretation of the geologic history of the islands. These rocks were formed by crystallization of igneous rocks from molten magma and by metamorphic recrystallization of ancient rocks as a result of high heat and pressure far below the surface and have subsequently been uplifted and exposed at the surface by erosion of the overlying rocks.

The Turtleback Complex includes several different kinds of crystalline rock. The oldest consists of dark, coarse–grained gabbro, which has been intruded by light colored diorite. Gabbro and diorite make up about two thirds of the rocks in the complex. Light colored granitic rocks make up about a fourth of the Turtleback Complex and occur as large bodies on the western part of Orcas Island and as smaller irregular bodies intruded into the older, dark, crystalline rocks. Both the granite and the gabbro were later intruded by smaller dikes and other small intrusions. Granite with large crystals occurs on Orcas, Jones, and Barren Islands and along the west shore of Deer Harbor.

Figure 17. Granitic rocks of the younger phase of the Turtleback Basement Complex at the public dock, Deer Harbor, Orcas Island.

The Turtleback basement complex has been affected by two stages of low temperature, high pressure metamorphism, which locally obscures primary igneous molten features. The earlier phase of metamorphism is characterized by recrystallization under high pressure, low temperature conditions. This metamorphism has not affected the overlying Devonian and younger beds so the metamorphism must be older than the unaltered rocks. A younger phase of metamorphism is characterized by intense shearing and crystallization of low temperature, low pressure minerals that also affected upper Paleozoic and Mesozoic rocks.

Age of the Turtleback Crystalline Complex

The Turtleback basement complex is overlain by Devonian sandstone and limestone about 360–400 million years old. Cobbles and pebbles of Turtleback rocks occur in Devonian conglomerate on the west side of Orcas Island and on O'Neal Island, showing that the overlying sediments were deposited on top of the eroded Turtleback Complex and that the Turtleback must therefore be older than Devonian.

Dating of Turtleback rocks using isotopes reveals even older ages. The oldest isotope dates of the Turtleback Complex range in age from late Pre–Cambrian (older than 540 million years) to Devonian (360–420 million years old). The oldest dates, 554 ± 16 and 545 ± 16 million years ago, come from gabbro, the oldest rocks in the complex. A potassium–argon isotope date of 554 ± 16 million years was obtained from hornblende in gabbro pegmatite along the shore of East Sound south of Crescent Beach (Fig. 18).

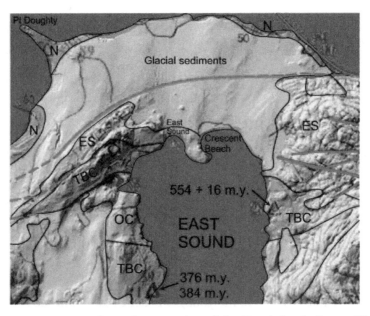

Figure 18. Isotope dates from rocks of the Turtleback Crystalline Complex (TBC), northern East Sound, Orcas Island. my=million years

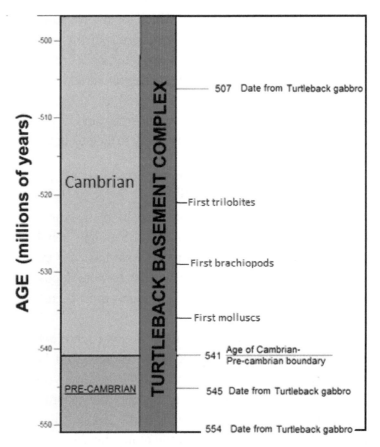

Figure 19. Age of the Turtleback Crystalline Basement Complex.

The age of the Pre–Cambrian–Paleozoic boundary has been placed at 540 million years by the International Commission on Stratigraphy. Because the 554 and 545 million year old dates are from intrusive gabbro, the time of crystallization is late Pre-Cambrian (Fig. 19). Younger granite in the Turtleback complex has been dated at 507, 460, 437, 409, 398, and 332 million years ago. These ages span the Cambrian, Ordovician, Silurian, and Devonian Periods of the Paleozoic Era. Thus, the Turtleback Crystalline Complex appears to span as much as 200 million years, from the Pre-Cambrian to the Devonian.

The best places to see these ancient rocks are on Orcas Island (Fig. 16) where they are exposed along sea cliffs and rocky ledges. Good examples may be seen at the West Sound marina, Turtleback Mt., along the shore of East Sound between Rosario Crescent Beach, and Moran State Park.

PALEOZOIC ERA
540 to 250 million years ago

The Paleozoic Era is a geologic time period that encompasses the earliest forms of life on Earth (Fig. 20). Its name comes from the Greek "Paleo," (meaning old or ancient) and "zoic" (meaning life). The Paleozoic Era spans the time from about 540 to 250 million years ago and was a time of intense geological and life–form changes. The beginning of the Paleozoic marked the dramatic expansion and diversification of life forms that had appeared late in the preceding Pre–Cambrian Era. Extensive, warm, shallow seas occupied much of North America and various other parts of the world. The first vertebrate organisms (e.g., fish, amphibians, and reptiles) appeared and marine organisms were plentiful (Fig. 20).

The Paleozoic Era is subdivided into seven geologic periods (the Cambrian, Ordovician, Silurian, Devonian, Mississippi, Pennsylvanian, and Permian), each containing unique fossils of former life in sedimentary beds. In the San Juan Islands, the Paleozoic rocks were formed by volcanic eruptions and deposition in former seas spanning the Devonian to Permian Periods.

ERA	PERIOD	LIFE FORMS	MILLIONS OF YEARS
PALEOZOIC	PERMIAN	Many organisms become extinct at the end of the Permian. Trilobites become extinct. Marine life flourishes, abundant molluscs, corals, brachiopods, fusilinid foraminifera	250
	PENNSYLVANIAN	Vast coal-making forests, first reptiles. Abundant corals, brachiopods, fusilinid foraminifera,	300
	MISSISSIPPIAN	Vast·coal-making swamps, first land vertebrates. Abundant corals, crinoids, bryozoans, brachiopods	325 360
	DEVONIAN	First amphibians, trees. Corals, crinoids, brachiopods, trilobites abundant.	420
	SILURIAN	Abundant corals, brachiopods, crinoids, trilobites. First·jawed·fish.	445
	ORDOVICIAN	Graptolites, crinoids, nautiloids, early·corals, trilobites, brachiopods	485
	CAMBRIAN	Major diversification of life forms. Trilobites, brachiopods.	540

Figure 20. Geologic time scale and life forms of the Paleozoic Era.

EAST SOUND GROUP
Devonian to Permian
420 to 250 million years ago

The Devonian Period, named after Devon, England where rocks from this period were first studied, spans the time from about 420 to 360 million years ago (Figures 13, 20). The Earth's continents looked completely different than they do now (you wouldn't even recognize North America, much less the San Juan Islands). Sharks and other fish were the dominant vertebrates in the sea, but on land, vertebrate animals had only begun to evolve. There were no mammals, only early primitive reptiles. Marine life was characterized by many kinds of abundant brachiopods (molluscs), corals, bryozoans, lily-like crinoids, and trilobites were common. The first ammonites (nautiloids) appeared and dominated sea life in the Mesozoic.

The oldest sedimentary rocks in the San Juan Islands consist of middle Devonian to Permian (about 390 to 250 million years ago) volcanic rocks, limestone lenses, chert, argillite (shale), and sandstone on Orcas, San Juan, Shaw, Jones, and a few of the smaller islands. These rocks make up the East Sound Group, which consists mostly of volcanic rocks, including much tuff and breccia (rocks composed of fragmental material blown out of volcanoes). Limestone interbedded with volcanic rocks contains middle Devonian (360-420 million years old) fossil corals, molluscs (brachiopods), conodonts (microscopic organisms with silica skeletons), and other marine organisms. Danner (1957, 1966) also identified Permian fossils (250–300 million years old) in some of the limestone. In order to distinguish between the older and younger limestone, rocks containing Devonian fossils were named the President Channel Formation by Danner (1957, 1966) (Fig. 21). The President Channel Formation consists of limestone lenses interbedded with altered volcanic rocks, chert, siltstone, greywacke sandstone, and conglomerate (Figs. 22, 23, 24), which are exposed in sea cliffs along the western margin of the Turtleback Range on Orcas Island. About half of these rocks consist of dense, green, altered volcanic rocks. Greywacke sandstone composed of volcanic fragments is also abundant. The total thickness of the President Channel Formation is about 1,400 feet.

The base of a 35–foot thick conglomerate bed exposed along the shore near Orcas Knob is made up of well-rounded, pebbles, cobbles, and boulders of granite, altered andesite, and quartzite that were most likely derived by erosion of the underlying Turtleback Basement Complex. Some of the granite boulders are up to two feet in diameter. The conglomerate is overlain by 25 to 50 feet of thin, alternating beds of sandstone and argillite, which are in turn overlain by a 30-foot-thick limestone bed containing Devonian fossils (Fig. 22). Contact relations on O'Neal Island also suggest a major discordance between the Devonian beds and the underlying crystalline rocks of the Turtleback Basement Complex

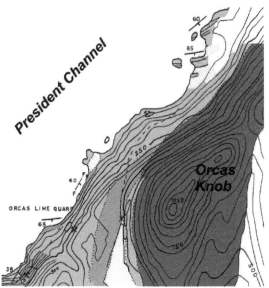

Figure 21. Geologic map of Orcas Knob, Orcas Island, showing the location of Devonian fossil-bearing limestone lenses (bright blue) interbedded with volcanic rocks (green) and greywacke sandstone (brown). The Devonian rocks are overlain by younger, late Paleozoic rocks (blue). (Modified from Danner 1957)

Figure 22. Fossil stromatoporoid *Amphipora* in Devonian rocks
at a beach west of Orcas Knob. (Photo by Ted Danner).

Figure 23. Devonian limestone lens interbedded with contorted chert and
argillite, West Sound quarry, Orcas Island. (Photo by Ted Danner)

Figure 24. Devonian limestone lens in argillite at White Beach
Bay, Orcas Island. (Photo by Ted Danner).

Fossils in the President Channel limestone include the brachiopod *Atrya* (Fig. 25), a mollusc typical of the late middle Devonian (about 390 million years ago). Also present are corals (Fig. 26), molluscs (brachiopods), gastropods (snails), and conodonts (microscopic fossils). Isotope dating of zircon in greywacke sandstone indicates an age younger than 300–400 million years, consistent with the Devonian age shown by fossils.

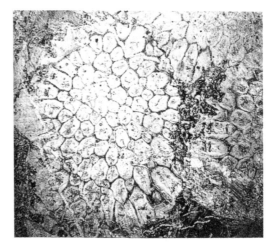

Figure 25. Devonian fossil mollusk,

Atrypa reticularis. (Photo by Ted Danner)

Figure 26 Devonian coral from limestone in
Red Cross quarry, Orcas Island. (Photo by Ted Danner)

26

The Permian Period (250–300 million years ago) was the last period of the Paleozoic Era. It was notable because of the flourishing of coral reefs, foraminifera, and brachiopods (molluscs). The end of the Permian was marked by the extinction of many marine species.

Permian sedimentary and volcanic rocks are widespread on Orcas, San Juan, and Henry Islands and less extensively on several other islands. Danner (1966) recognized several Permian geologic units, including (1) middle Permian basaltic submarine volcanic rocks, (2) middle Permian ribbon chert, and (3) limestone.

The oldest Permian rocks on San Juan Island consist of basaltic pillow lava, breccia and tuff, ribbon chert, and limestone exposed for about two miles along the west coast. Pillow lavas are submarine lava flows made up of bulbous ellipsoidal masses of lava about the size of pillows.

Both limestone and ribbon chert are interbedded with the volcanic rocks. The limestone occurs mostly as small lenses, the largest being about 1,500 feet long and 300 feet wide. Limestone also appears as angular fragments up to several feet in diameter in the volcanic breccia. The limestone is fine grained and contains well preserved fossils, mostly microscopic shells of fusulinid foraminifera (Fig. 27), and calcareous algae, which suggest shallow water deposition, probably on reefs fringing volcanic islands or on shallow offshore carbonate banks.

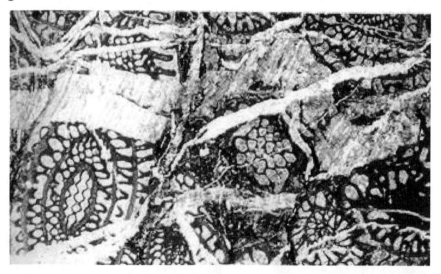

Figure 27 Microscopic Permian fusulinids in limestone. (Photo by Ted Danner)

Ribbon chert is interbedded in the volcanic rocks as lenses a few feet to a hundred feet thick. The chert occurs as beds about an inch thick, separated by thin shale partings. The chert contains remains of siliceous marine microorganisms known as radiolarians. The association of the chert with shallow water limestone suggests that the chert is also of shallow marine origin. Some interbedded coarse grained sandstone is made up entirely of small, angular, chert fragments.

Limestone pods and lenses interbedded with pillow basalt and volcanic breccia occur at Lime Kiln Point on western San Juan Island (Fig. 28, 29). The limestone contains fusulinids that indicate a Permian age. Radiolarian and conodont microfossils, which range in age from Permian to Triassic, are also present.

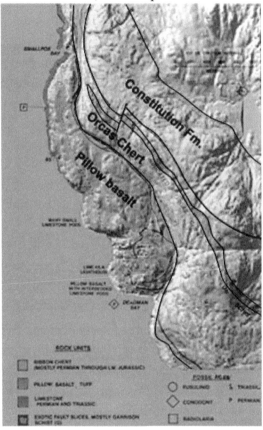

Figure 28. Geologic map of Lime Kiln Point. (Modified from Brandon et al., 1988)

Figure 29. Limestone lens in volcanic rocks, Lime Kiln Point, western San Juan Island.

GARRISON SCHIST

Well–foliated greenschist, amphibolite, phyllite, and mica–rich quartzite occur at Garrison Bay and near Rosario at East Sound on Orcas Island. Up to 400 feet of the schist crops out as a discontinuous body over a broad area on northwestern San Juan Island where it extends for about eight miles with only one major break.

Garrison phyllitic quartzite formed from recrystallization of thin interbeds of chert and argillite where the chert was heated to form quartzite and the argillite was recrystallized to mica. The greenschist and amphibolite formed from metamorphic recrystallization of basaltic lava under low temperature and high pressure. Amphibolite occurs with the greenschist but appears to be restricted to northwest San Juan Island near Garrison Bay and eastern Orcas Island.

Garrison schist occurs as fault slices 3 to 12 feet thick at or near the contact between Orcas Chert and the overlying Constitution Formation. The Constitution Formation locally contains angular pebbles of the schist, indicating that the schist must be older than the Constitution formation.

Two potassium–argon isotope dates, 286 ± 20 and 242 ± 14 million years, from Garrison amphibolite indicate a Permian to early Triassic metamorphic age. A potassium–argon isotope age determination on amphibole from Garrison greenschist yielded a date of 167 ± 12 million years (late Jurassic).

Other Permian Rocks. Permian fusulinid fossils occur in two limestone lenses at Double Hill north of East Sound on Orcas Island. The limestone, interbedded with volcanic rocks and greywacke sandstone, is the oldest Permian rock in the Islands.

At Judd Cove in upper East Sound on Orcas Island, limestone lenses containing late Permian fusulinid fossils are interbedded with volcanic rocks, ribbon chert, sandstone, and breccia made of chert fragments. These rocks occur as fault slices in crystalline rocks of the Turtleback Complex.

MESOZOIC ERA
(250 to 65 million years ago)

The Mesozoic Era is referred to as the age of dinosaurs and reptiles because of the flourishing of these creatures from about 250 million years ago to 65 million years ago. During this time, dinosaurs and other reptiles dominated the land and ammonites and belemnites (nautiloids) abounded in the global seas. The super continent of Gondwana broke apart, splitting it into two large continents in the Jurassic Period (150–200 million years ago). In the late Cretaceous Period (about 65-90 million years ago), the restless Earth went on a bit of global tectonic rampage with intense folding, faulting, and mountain building in most of world's alpine areas. This period of crustal deformation was also felt in the San Juan Islands.

ERA	PERIOD	LIFE·FORMS·AND·EVENTS	MILLIONS OF·YRS
MESOZOIC	**CRETACEOUS**	Laramide mt. building, intense deformation of rocks in San Juan Islands, Cascades, and all over the world. Dinosaurs dominate land, then become extinct during great global extinction. Ammonites become extinct.	65
	JURASSIC	Many species of dinosaurs. First birds. Ammonites flourish in the sea.	150
	TRIASSIC	Ammonites abundant, first mammals. Dinosaurs appear	200 250

Figure 30. Mesozoic geologic time scale.

Orcas Chert
Triassic–early Jurassic
250 to150 million years ago

The middle Permian (250 to 300 million years ago) volcanic rocks on San Juan Island are overlain by thick, highly contorted Orcas ribbon chert (Figs. 31, 32) with minor interbeds of basaltic tuff, pillow basalt, and limestone lenses. The ribbon chert is so called because it consists of multiple layers 1-2 inches thick separated by thin argillite partings a fraction of an inch thick, producing a ribbon–like appearance (Figs. 31, 32). Much of the silica in the chert comes from silica skeletons of microscopic radiolarian fossils that accumulated on the sea floor.

Figure 31. Orcas ribbon Chert, Cattle Point, San Juan Island.

Figure 32. Contorted Orcas Chert, Orcas Island.

The chert is up to 1,500 feet thick on western San Juan Island and 2,500 feet thick on Orcas Island. Radiolaria and conodonts (microscopic fossils of silica) in ribbon chert from 27 localities indicate that the Orcas Chert ranges in age from Triassic (200–250 million years ago) to early Jurassic (150–200 million years ago). Limestone on northern San Juan Island includes lenses of Orcas ribbon chert that contain late Triassic (200–250 million years ago) radiolarian microfossils.

The volcanic rocks that occur beneath Orcas Chert along the west coast of San Juan Island are absent on Orcas and northern San Juan Island so the chert lies directly on Pre–Cambrian rocks of the Turtleback Crystalline Basement Complex. Bedding in the chert is parallel to the contact with the Turtleback crystalline rocks, suggesting that the chert was deposited on the eroded surface of the Turtleback. Thick, sheared shale and greywacke sandstone of the Constitution Formation were deposited on top of Orcas Chert on San Juan, Shaw, and Orcas Islands. The contact between Orcas chert the overlying Constitution greywacke sandstone is exposed along the northeast shore of San Juan Island about a mile east of Sportsmans Lake.

Orcas Chert makes up the northwest end of the San Juan Island, Mount Dallas, and the area northwest of the San Juan Range. Chert along the southwest coast of San Juan Island trends approximately parallel to the coast and dips to the northeast. Along the northeast coast of the island, the chert dips towards the southwest. The chert contains lenses of limestone, the largest of which is at Roche Harbor (Fig. 33) where it has been extensively quarried.

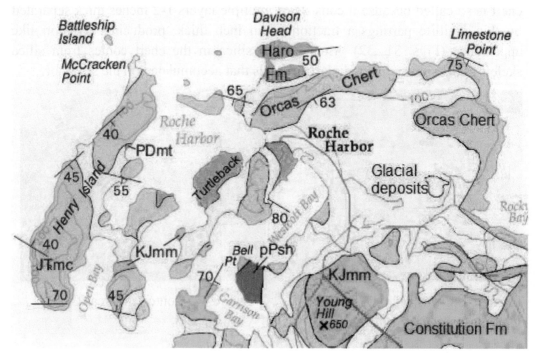

Figure 33. Extent of Orcas Chert on geologic map of northern San Juan Island.

31

CARTER POINT—LUMMI FORMATION
Late Jurassic to early Cretaceous
160 to 100 million years ago

At Carter Point on southern Lummi Island (Figs. 34, 35), volcanic-rich greywacke sandstone, chert, and basalt several thousand feet thick lie on chert and pillow basalt, which may (or may not) be the Orcas Chert. Calkin (1959) named these rocks the Carter Point Formation, but it was later renamed the Lummi Formation. The term Lummi Formation was subsequently used for some rocks in the San Juan Islands that are clearly not correlative, so because Carter Point Formation has precedence and to make it clear that the Lummi Formation discussed here is restricted to Lummi Island, the term Carter Point–Lummi Formation is used.

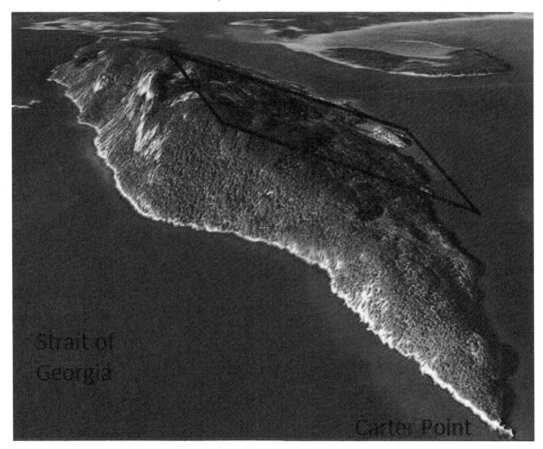

Figure 34. Lummi Island looking north. Thick greywacke sandstone has been tilted to the northeast (right). The more gentle slopes on the eastern slope are essentially parallel to the bedding of the Carter Point–Lummi Formation (tilted square).

Almost the entire high, southern portion of Lummi Island (Figs. 34, 35) is made up of highly indurated greywacke sandstone (Fig. 36), argillite (Fig. 37), and conglomerate several thousand feet thick. Massive beds of conglomerate 150-200 feet thick form high, vertical cliffs along the southwest side of the island.

Figure 35. Topographic map of Lummi Island.

The greywacke sandstone is made up almost entirely of volcanic, argillite, and chert sand grains (Fig. 36). Graded bedding (Fig. 38) is common in both the sandstone and conglomerate. The rocks are cut by numerous quartz veins (Fig. 39). The greywacke is extremely well indurated and thus, quite resistant to erosion. Because of this, these rocks have been extensively quarried for rip-rap and crushed for road building.

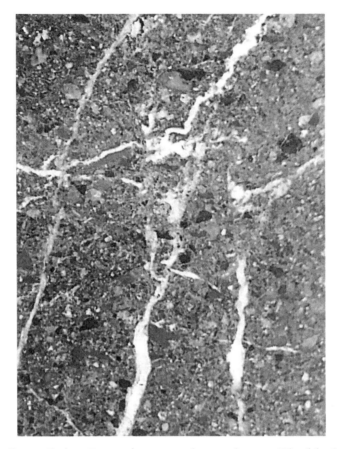

Figure 36. Carter Point–Lummi greywacke sandstone. The black grains are argillite, the white grains are chert, and the rest is mostly green metamorphosed volcanic grains. Most of the grains are not well rounded and are poorly sorted.

Figure 37. Carter Point–Lummi argillite.

Figure 38. Graded bedding in greywacke sandstone. Each graded bed consists of lower, light, coarse–grained sediment grading upward into finer and finer particles, eventually becoming dark colored.

Figure 39. Carter Point–Lummi Greywacke sandstone cut by many quartz veins.

The Carter Point–Lummi Formation beds on southern Lummi Island dip more or less uniformly to the east (Fig. 40) and are not tightly folded. A stratigraphic section measured by Calkin (1959) on the southwest side of the island shows more than 1000 feet of sandstone, shale, and conglomerate (Fig. 41). The lower ~800 feet consist dominantly of sandstone and shale with very little conglomerate. The upper 300 feet of the section has many conglomerate beds.

Two isotope ages have been obtained from rocks of the Carter Point–Lummi Formation on Lummi Island. A potassium–argon isotope date of 160 ± 22 million years was from metamorphosed basalt and a fission–track date of 107 ± 13 million years was obtained from greywacke sandstone. Radiolarian microfossils from five localities on Lummi Island indicate a late Jurassic to early Cretaceous age, consistent with both the 107 and 160 million year isotope dates.

The Carter Point–Lummi Formation is a geologic enigma—it consists of greywacke sandstone, chert, and basalt which have been metamorphosed under high pressure, low temperature conditions, similar to the late Paleozoic-early Mesozoic rocks in the islands. However, isotope dating and fossil ages suggest that the Carter Point–Lummi Formation is roughly the same age as the Haro, Spieden, and Deer Point sandstones, which are relatively unmetamorphosed and lack interbedded chert and pillow basalts commonly found in the Carter Point–Lummi and Constitution Formations. So the question here is, if the Haro, Spieden, and Deer Point sandstones and Carter Point–Lummi and Constitution Formations are the same age, why are they not interbedded with chert and pillow basalt and show the same degree of metamorphism? Several possibilities exist for these anomalies:

(1) The occurrence of chert and pillow basalt in the Lummi–Carter Point and Constitution Formations might have been due to different depositional environments than the Haro, Spieden, and Deer Point Formations, and their metamorphism might have been due to much deeper burial than the unmetamorphosed rocks.

(2) The Lummi–Carter Point Formation might have been formed elsewhere and carried to its present position by continental drift.

Which of these possibilities is correct is not readily apparent at this point, but what is clear is that the Carter Point–Lummi and Constitution Formations are significantly different than rocks of similar age in the islands.

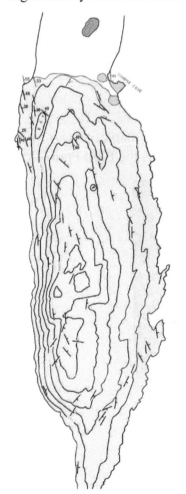

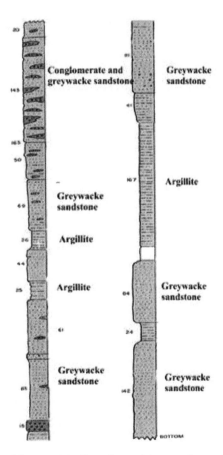

Figure 40. Eastward–dipping Carter Point–Lummi Fm., Lummi Island. (Modified from Calkin, 1959)

Figure 41. Stratigraphic section of the Carter Point–Lummi Fm. (Modified from Calkin, 1959)

HARO FORMATION
(Upper Triassic, 200–250 million years ago)

Davidson Head, a peninsula at the northern end of San Juan Island connected to the main island by a narrow gravel bar (Figs. 42, 43), is composed of conglomerate (Fig. 44), shale, greywacke sandstone, and limestone known as the Haro Formation of upper Triassic age (200–250 million years ago). The lowermost beds consist of 920 feet of conglomerate with boulders of Orcas chert, greywacke sandstone, granite, porphyry, andesite, and limestone.

Figure 42. LIDAR image of Davidson Head, northern San Juan Island.

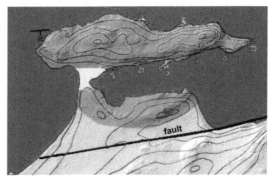

Figure 43. Geologic map of Davidson Head. Brown area is the Haro Formation.

Figure 44. Haro Formation conglomerate along the south shore of Davidson Head, northern San Juan Island. The large white cobble is a granite. Most of the other pebbles are volcanic. (Photo by Ted Danner)

The conglomerate is overlain by 330 feet of thin-bedded shale, greywacke sandstone, and limestone (Figs. 45, 46). The volcanic composition and texture of the Haro rocks suggests that they were derived from a nearby active volcanic source and deposited in a volcanic island arc environment. Unlike the Paleozoic rocks of the San Juan Islands, which are highly sheared and show low grade metamorphic recrystallization, the Haro sediments are relatively fresh and unaltered.

Figure 45. Vertically tilted sandstone and shale of the Triassic Haro Fm. along the south shore of Davidson Head. (Photo by Ted Danner)

Figure 46. Thin beds of sandstone and shale of the Haro Formation along the western shore of Davidson Head, San Juan Island. (Photo by Ted Danner)

The limestone beds are less than four feet thick and interbedded with shale. In places, limestone grades into breccia and conglomerate with sharp contacts. The limestone, best exposed along the eastern part of the head, is composed almost entirely of thin, compressed and flattened shells of the clam *Halobia* (Figs. 47, 48). Fossils are best preserved in organic shale just below the limestone. Many of the shells have been replaced by pyrite. The purer limestone is laminated and contains thin interbeds of gray chert. The *Halobia* fossils indicate an upper Triassic age (200–250 million years) for the Haro Formation.

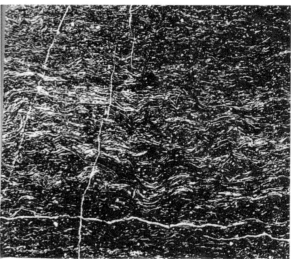

Figure 47 Upper Triassic clam *Halobia*, Davidson Head, San Juan Island. (Photo by Ted Danner)

Figure 48. Haro Formation limestone made up entirely of flattened and compressed *Halobia* shells. (Photo by Ted Danner)

CONSTITUTION FORMATION

The Constitution Formation is important because it is the most widespread rock unit in the San Juan Islands. Thick greywacke sandstone, siltstone, ribbon chert, and minor lava overlie the Orcas Chert on parts of Orcas, San Juan, and Shaw Islands (Figure 49).

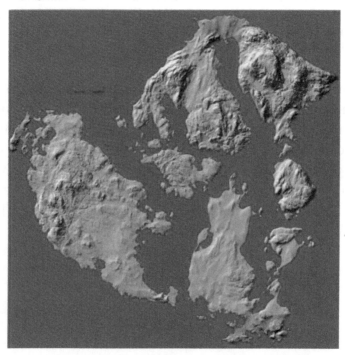

Figure 49. Extent of the Constitution Formation (green).

The name Constitution Formation was first used for rocks on eastern Orcas Island (italicized text) (Easterbrook, 1958).

Constitution Formation

"The argillite and greywacke of the Constitution Formation lie above the Chert Formation and crop out on the eastern slope of Entrance Mt., Rosario Hill, Mt. Pickett, and Mt. Constitution (from which the name of the formation is taken). Along the eastern slopes of Entrance Mt. and north of Rosario, the strikes and dips of the Constitution Formation parallel those of the Chert Formation, suggesting that it is conformable on the chert. However, above Cascade Lake on the western slope of Mt Constitution, a relatively small thickness of chert with associated limestone lenses is cleanly truncated by the Constitution greywacke. The chert abuts against the contact at a high angle, and with no evidence of faulting, such as slickensides, gouge, drag folding, or secondary quartz along the contact is seen, the contact seems to be an unconformity." (Easterbrook, 1958)

 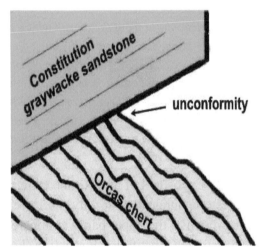

Figure 50. Constitution greywacke sandstone lying across the upturned beds of Orcas chert, eastern Orcas Island above Cascade Lake. The discordant bedding shows that sometime after the chert was deposited, the beds were tilted and eroded to a flat surface before deposition of the overlying Constitution greywacke sandstone.

"Near the base of the sequence along the coast near Olga, the argillite and greywacke is highly siliceous and has been intensely sheared and contorted. Black argillites interbedded with thin layers of chert from 1/8 to 1/4 inch in thickness have been intricately folded and contorted. Associated with these rocks are grayish colored greywackes, which have also been intensely sheared. In many places the greywacke is entirely enclosed by black argillite and vice versa, giving the appearance that the two have been

40

scrambled together by movements along innumerable, tiny, shear planes. An idea of the scale of the shearing can be seen where many small quartz veinlets have been offset into numerous stairstep–like patterns. In the field it is obvious that many of these 'inclusions' of greywacke within argillite are not depositional features.

The argillites near the base of the section are thinly interbedded with layers of white to gray colored chert, but the overall color of the rock is dominated by the black argillite. The chert layers average about 1/8 inch in thickness, occurring generally about every quarter of an inch. Because they are more resistant to weathering than the argillite they stand out as small protruding ribs. For the most part, the rocks near the base are highly siliceous and in many places have been shot full of quart veinlets .

Along the southeast shore from Sea Acres to Doe Island, the argillites are also highly siliceous and small thicknesses of banded blue-gray ribbon chert similar to those of the Chert formation are found. Associated with these rocks are dense, hard, green volcanics and some breccia. South of Doe Bay a few minor lenses of limestone only a few feet thick were found, but no fossils were seen.

Farther inland from the coasts where the rocks are presumably higher up in the sequence, the argillite becomes less siliceous and greywacke becomes more common. The argillite is not nearly as hard as those near the base of the sequence and is more easily weathered. The thin interbedded layers of chert are not present and consequently, bedding is not often seen in the outcrops. The greywacke is usually massive and generally well cemented, although not as hard as the siliceous greywacke near the base."
(Easterbrook, 1958)

Nearby exposures of the Turtleback crystalline complex beneath the chert and limestone suggests that only the lower part of the Orcas chert is present here and much of the upper part was apparently eroded away before deposition of the overlying Constitution Formation (Fig. 50). Along the NE coast of Orcas Island, beds of the Orcas chert and Constitution Formation appear to be essentially parallel, rather than at a steep angle to one another.

Figure 51 Constitution Formation siltstone (at hammer) lying on Orcas chert and overlain by greywacke sandstone. East Sound north of Rosario.

Constitution greywacke sandstone overlies Orcas chert along the west sides of Mt. Entrance and Mt. Constitution (Fig. 51) where they trend about N35°–45°E and dip to the southwest. Greywacke sandstone and siltstone along the northeast shore from Point Lawrence to the foot of Buck Mountain trend about N60°W and dip 60°SW. Orcas chert and Constitution Formation on Mt. Constitution are clearly in fault contact as their bedding is nearly at right angles to one another.

A precipitous cliff along the east side of Mt. Constitution follows a large, high–angle fault (Fig. 52). The scarp drops from the summit of Mt. Constitution (2409') to Mountain Lake (916') on the down–dropped block to the east, a vertical offset of about 500 feet. This fault separates Constitution greywacke sandstone, which makes up Mt. Constitution, from rocks of the Turtleback Crystalline Complex to the east. The southern margin of the Constitution Fm. is also a major fault contact. The Olga fault (Fig. 52), a major fault that follows the valley connecting the villages of Olga and Doe Bay, separates the Constitution Fm. from the Deer Point Fm. to the south.

Rocks of the Constitution Formation make up most of San Juan and Shaw Islands and the southern part of Orcas Island between East Sound and West Sound (Fig. 49). The contact between Constitution greywacke sandstone and the underlying Orcas chert is exposed along the northeast shore of San Juan Island about one mile east of Sportsmans Lake and on the flanks of Mt. Constitution.

Figure 52. LIDAR image showing the high–angle fault on the east flank of Mt. Constitution, separating Constitution greywacke sandstone from rocks of the Turtleback Crystalline Complex to the east.

The maximum thickness of the Constitution Fm. occurs on San Juan Island where McLellan (1927) described several depositional units.

"The lowest member ...., as exposed on San Juan Island, is a tuffaceous and cherty greywacke. It overlies the Orcas cherts with approximate conformity. This greywacke is typically fine-grained and occurs in beds of varying thickness. In places it appears as thin beds which differ from each other only in the size of the fragments composing them. Occasionally the greywacke contains scattered layers of chert, and it is possible that the conditions which brought about the deposition of the Orcas cherts continued on, and additional sediments, including tuffaceous material, were added to them. The basal greywackes have a probable maximum thickness of several thousand feet.

The basal greywacke outcrops throughout the elevated region in the vicinity of Point Caution, and on account of their resistance to erosion, the outcrops form a series of elevated ridges." "They are seen on Biological Hill, the San Juan Range, Mount Grant, and on Little Mountain. The upper limit of the greywackes is marked by a well defined formation about 100 feet thick, composed of coarse breccia and conglomerate. This formation underlies a portion of the village of Friday Harbor and is encountered at several localities near the southwestern side of San Juan Island. The breccia formation is followed by several hundred feet of carbonaceous argillite which has no banding or apparent structure of any kind. This is followed by a tremendous thickness of thin-bedded alternating light and dark layers of argillite which often grades into phyllite. Occasionally there are interbeds of greywacke, grit, or conglomerate. The uppermost strata of the group are composed of thick conglomerate, grit, greywacke, and slate." (McLellan, 1927)

Most of the Constitution Fm. is massive with very little recognizable bedding so the total thickness of the rocks is difficult to measure, but appears to exceed 10,000 feet on San Juan and Orcas Islands. The Constitution Fm. consists of three members: (1) a lower member composed of black siltstone, green tuff, ribbon chert, minor greywacke sandstone, pillow basalt, and small lenses of limestone, (2) a middle member composed of massive greywacke sandstone with minor conglomerate, and (3) an upper member composed mostly of thin–bedded black siltstone and greywacke sandstone.

Garrison schist is intricately thrust faulted in slices at the base of the Constitution Fm. and perhaps within the underlying Orcas chert. This zone is near the stratigraphic transition between the Orcas chert and overlying Constitution Fm. The middle and upper units of the Constitution Fm. are free of these fault slices. The upper member contains about 30 percent thin, tabular, pillow lava 10 to 45 m thick. The chemistry of these lavas suggests that they were erupted on the ocean floor.

McLellan (1927), Danner (1957, 1966), Easterbrook (1958), and Vance (1975) all recognized that the Constitution Fm. was deposited directly upon the underlying Orcas chert on the basis of the depositional contact. Conglomerate, containing well-rounded pebbles of Turtleback crystalline rocks and chert, overlies steeply dipping, truncated Orcas chert (1) on Orcas Island at the southeast end of Cascade Lake, (2) at the small bay just east of the town of Orcas, and (3) on northwestern San Juan Island, where 100-foot-thick beds of basal conglomerate occur widely at the contact of the Orcas chert and the Constitution Fm. Conglomerate, composed of Turtleback granite, Orcas chert, limestone, and Garrison schist, is exposed south of Garrison Bay above the Orcas chert at the base of the Constitution Fm. Orcas

The Constitution Fm. is highly deformed and contains thrust–faulted slices of Turtleback granite and Garrison schist. It also shows intense slickensides and internal shearing deformation, accompanied by widespread, low temperature metamorphism, which resulted in small crystals of the metamorphic mineral prehnite in numerous white veins.

The large thickness of Constitution greywacke sandstone, composed mostly of sand and silt grains of volcanic origin, suggests deposition at a continental margin near a volcanic arc, perhaps similar to the modern day volcanic arcs offshore from the Aleutian or the Japanese chains of volcanic mountains.

Age

Radiolarian fossils (siliceous microfossils) in chert beds within the Constitution Fm. suggest a late Jurassic to early Cretaceous age. Isotope dates from zircon in the Constitution Fm. indicate an age younger than 150 million years, which would indicate deposition about at the Jurassic/Cretaceous boundary.

SPIEDEN FORMATION
Late Jurassic to early Cretaceous
(~160 to 100 million years ago)

Conglomerate, sandstone, shale, and limestone of the Spieden Formation occur on Spieden Island, Sentinel Island, and Sentinel Rock (Figs. 53, 54). Conglomerates (Fig. 55) makes up about 85 per cent of the rocks. They contain cobbles up to a foot in diameter, but most are not larger than one inch. Pebbles and cobbles are composed of andesite, diorite, granite, porphyry, quartz, jasper, greywacke, chert, argillite, and limestone. Rare boulders of fine grained granite also occur. Conglomerate on Sentinel Island has the same composition as that on Spieden Island.

Figure 53. Spieden, Sentinel, and Cactus Islands.

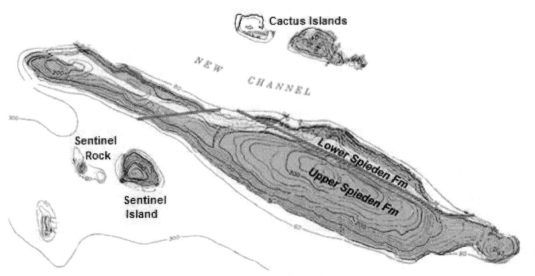

Figure 54. Geologic map of Spieden and Sentinel Islands.

Figure 55. Early Cretaceous conglomerate, southern shore of Spieden Island.

Along the northern margin of Spieden Island, about 35 feet of thin bedded shale contains many fossils and grades upward into sandstone with occasional beds of muddy limestone. The fossil–bearing beds are overlain by more than 2,000 feet of conglomerate that trends N 70–75°W, and dip 45–60°S. At the foot of Spieden Bluff, the fossiliferous shale and sandstone strike N 65° W, and dip 65° SW.

The Spieden Formation and has been broken into two subunits, the Spieden Bluff Member (older), and the Sentinel Island Member (younger) (Fig. 56). The lowermost Spieden Bluff Member is composed of 265 feet of volcanic breccia and conglomerate with minor sandstone and shale. The upper part of the unit consists of 65 feet of sandstone and shale containing late Jurassic marine fossils, primarily the clam *Buchia*. The Sentinal Island Member consists of 460 feet of thinly bedded sandstone and shale containing early Cretaceous marine fossils.

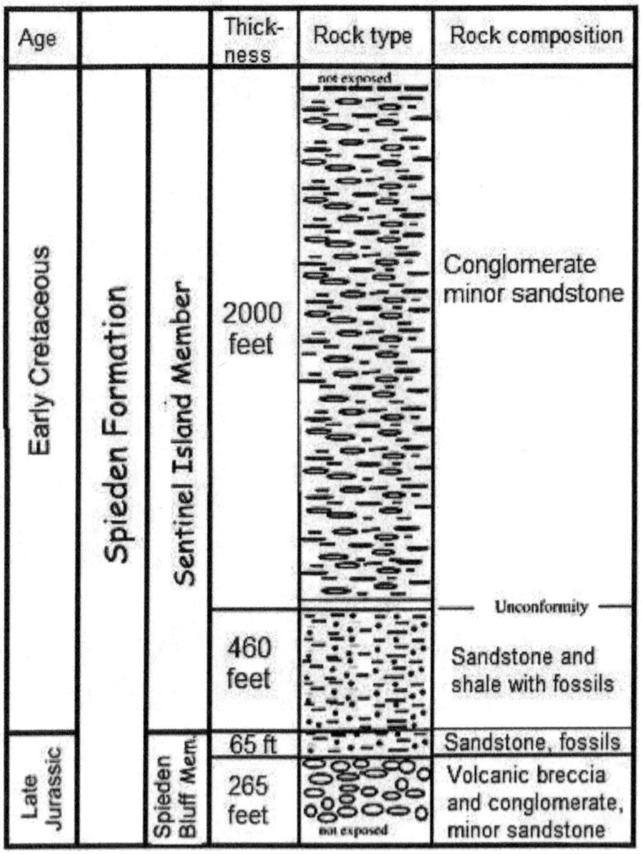

Age			Thick-ness	Rock type	Rock composition
Early Cretaceous	Spieden Formation	Sentinel Island Member	2000 feet		Conglomerate minor sandstone
					Unconformity
			460 feet		Sandstone and shale with fossils
			65 ft		Sandstone, fossils
Late Jurassic		Spieden Bluff Mem.	265 feet		Volcanic breccia and conglomerate, minor sandstone

Figure 56. Stratigraphic section of rocks on Spieden Island.

Fossils

The marine sandstone and shale on Spieden Island locally contain abundant fossils, mostly in the lower, older beds. The oldest fossils occur in the late Jurassic Spieden Bluff Formation (Fig. 56), a 65–foot–thick bed of sandstone and shale overlying 265 feet of volcanic breccia and conglomerate. The most common fossils are clams (*Buchia*).

The Jurassic beds are overlain by 460 feet of fossil-bearing sandstone and shale of the Sentinel Island Formation. Fossils are common on the north shore of Spieden Island where beds of fossil-bearing sandstone and shale of lower Cretaceous age occur. The clam *Aucella crassicollis* (Fig. 57) is the most abundant fossil, making up about 95% of the fossils found on the island. Among the fossils found on the island are the following:

Aucella crassicollis (clam)
Inoceramus (clam)
Belemnites (nautiloid)
Serpula (worm tube)

47

Figure 57 Fossil clam shells (*Aucella crassicollis*) from the north shore of Spieden Island. (From McClellan, 1927)

The lower fossil-beds are overlain by 2000 feet of coarse conglomerate that doesn't contain fossils, but is older than the late Cretaceous Nanaimo sandstone on Stuart Island just to the north and younger than the underlying late Jurassic beds. Isotope dating of zircon in the Spieden Fm. indicates ages as young as 140-160 million years, consistent with the early Cretaceous age given by fossils.

DEER POINT FORMATION
Cretaceous (100 million years ago)

Well–bedded conglomerate, greywacke sandstone, and shale are exposed on Orcas Island on the peninsula between Olga and Obstruction Pass. Beds of greywacke sandstone and shale are interlayered with the well–bedded siltstone and shale (Figs. 58–60). Massive greywacke conglomerate beds contain pebbles of chert, interbedded with thin-bedded shale.

Rocks of the Deer Point Fm. are isolated from older rocks to the north by the Olga fault, a major fault across southern Orcas Island and Peavine Pass to the south. The Deer Point beds have been moderately folded and faulted but are not nearly as deformed or recrystallized as the older rocks on Orcas Island (Figs. 58-60). They are not cut by igneous intrusions of any kind, lack quartz veins, and chert and volcanic rocks are totally absent, so they were mapped as the Deer Point Formation, a distinctly younger unit than other rocks on the island (Easterbrook, 1958). Other geologists have included the rocks on the Deer Point peninsula within the Lummi Formation, but fossils and isotope dates confirm that the Lummi Formation is older. Radiolarian microfossils in chert of the Lummi Formation at six localities on Lummi Island all indicate a Jurassic age (150–200 million years) and an isotope date of 160 ± 22 million years from a basalt flow on Lummi Islands confirms the Jurassic age of the Lummi Formation. An isotope date of 107 ± 15 million years was obtained from zircon in greywacke sandstone, indicating that the Deer Point Formation is Cretaceous in age and definitely younger than the Lummi Formation.

48

Figure 58. Deer Point greywacke sandstone south of Olga, Orcas Island.

Figure 59. Deer Point greywacke sandstone tilted to near vertical, south of Olga.

Figure 60. Thin-bedded Deer Point shale.

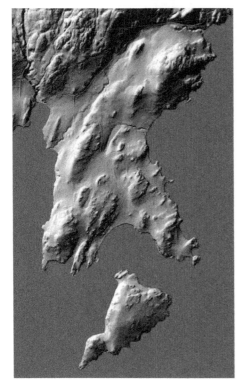

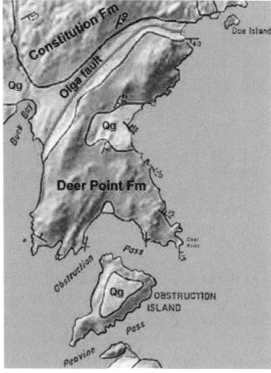

Figure 61. Lidar image of the Deer Point peninsula and Obstruction Island

Figure 62. Geologic map of the Deer Point Peninsula and Obstruction Island. Qg=Ice Age glacial sediments.

NANAIMO GROUP
Upper Cretaceous (65 to 105 million years ago)

Much of the northern San Juan Islands consists of sandstone, shale, and conglomerate of the upper Cretaceous Nanaimo Group (Figs. 63, 64), which, although faulted and folded, are much less deformed than the older rocks of the central San Juan Islands (Orcas, San Juan, Lopez, and Shaw). The Nanaimo beds were originally deposited as sand, mud, and gravel on the floor of an ancient sea that extended from the eastern San Juan Islands to Vancouver Island during the last part of the Cretaceous Period about 60-90 million years ago. Shells of marine organisms living in the sea were deposited with the sediments and are now preserved in the rocks as fossils.

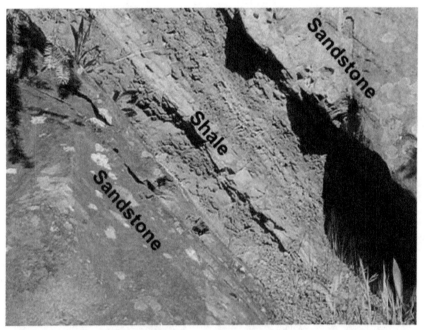

Figure 63. Dipping Nanaimo shale and sandstone, Sucia Island.
(Photo by George Mustoe)

As the depositional basin subsided, sediment derived from nearby granite hills was deposited in shallow marine water. The sediment consisted mostly of feldspar and quartz eroded from the granite and the resulting rock was arkosic sandstone, quite unlike older, volcanic–rich greywacke sandstones. Unlike the deep–water chert and pillow basalt of the older greywacke sandstones, no deep–water sediments occur in the Nanaimo sandstone. Interbedded conglomerate (Fig. 65), some containing boulders up to two feet in diameter on Orcas, Stuart, and Clark Islands, indicate that vigorous streams discharged their bedload of gravel into the shallow sea close to land.

Nanaimo sandstone on Orcas Island

The oldest Nanaimo beds on Orcas Island consist of 115 feet of massive sandstone interbeded with pebble conglomerate and overlain by 925 feet of fossiliferous shale. These beds are overlain by about 100 feet of distinctive pebble–cobble–boulder conglomerate that grades upward into pebble conglomerate interbedded with sandstone. Near the base of the conglomerate boulders a foot or more in diameter grade upward into interbedded pebble conglomerate and sandstone. Most of the pebbles and cobbles in the conglomerate consist of granite derived from the Turtleback Complex and chert derived from the Orcas chert. Fossil clams and ammonites occur above and below the conglomerate. About 485 feet of overlying shale interbedded with sandstone and pebble conglomerate contains fossil wood and leaves.

Figure 64. Dipping beds of Nanaimo sandstone on the north shore of Orcas Island. Joe Vance and Ross Ellis are at the far end of the outcrop.

Pebbles and cobbles in massive conglomerate at Point Doughty are composed of all of the older rocks exposed in the San Juan Islands, including boulders of coarse granite not seen in this region. Thin beds of conglomerate, sandstone, and organic shale rich in fossil leaf impressions overlie the Point Doughty conglomerate. Nanaimo beds on Orcas Island persistently trend N 65°W and dip to the south.

Figure 65. Nanaimo conglomerate at Pt. Doughty, northern Orcas Island.

Stratigraphic section, from Point Thompson southwestward:	Thickness (feet)
Coarse sandstone	30
Sandy shale with sandstone interbeds	776
Coarse sandstone	19
Thin-bedded sandstone and shale	20+
Concealed	1,745
Conglomerate with interbedded grit	50
Concealed	1,240
Coarse sandstone	30
Sandy shale with sandstone interbeds	1,050
Coarse conglomerate (Point Doughty)	100
Concretionary sandy shale	2
Coarse gritty sandstone	3
Carbonaceous sandstone, fossil plants	18
Concretionary shale	25
Fine-grained sandy conglomerate	10
Sandy shale	37
Coarse and fine conglomerate	65
Lignitic shale with fossil plants	400
Coarse sandstone	25
Total	5,645

Nanaimo sandstone on Stuart Island

Beds of Nanaimo sandstone, shale, and conglomerate make up Stuart Island and the smaller islands to the east. The Nanaimo beds on Stuart Island are 3,700 feet thick and consist of three members. The oldest member is made up of about 1,460 feet of thin–bedded sandstone, siltstone, and shale. Beds of massive sandstone, mostly less than 100 feet thick, and conglomerate beds a few tens of feet thick are also present. The lower member is overlain by about 1000 feet of conglomerate interbedded with cross-bedded coarse sandstone. Pebbles in the conglomerate consist of dense volcanic rocks, chert, siltstone and granite. About 240 feet of interbedded sandstone and shale containing fossil clams, brachiopods, and ammonites are the youngest beds exposed.

The Nanaimo beds on Stuart Island is made up of three units: (1) Pender sandstone and shale, (2) Extension conglomerate, and (3) Haslam shale and sandstone (Figs. 66, 67)

Pender Formation (uppermost unit) consists of ~500 feet of sandstone and
   shale containing marine fossils, mostly *Inoceramus*, which are abundant at
   Fossil Cove.

Extension conglomerate consists of 500 feet of massive conglomerate.

Haslam Formation (lowermost unit) consists of 1000 feet of shale with
   interbedded sandstone and conglomerate. A few fossils occur, but are rare.

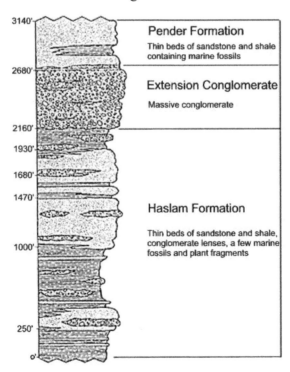

Figure 66. Composite stratigraphic section of rocks making up Stuart Island.

54

Figure 67. Steeply dipping Nanaimo thin–bedded shale, Stuart Island.

Nanaimo on Sucia

Sucia conglomerate

About 50 feet of bluish–gray conglomerate is exposed near the base of the Nanaimo section at the entrance to Fossil Bay on Sucia Island (Fig. 68). This unique conglomerate is composed of pebbles and cobbles of phyllite, white vein quartz, and greenschist, overlain by about 700 feet of richly fossiliferous shale, containing abundant shells of clams, ammonites, snails, nautiloids, and foraminifera. The Nanaimo beds on Sucia were divided into two stratigraphic units, the Sucia conglomerate and the Fossil Bay sandstone (Easterbrook, 1958).

Sucia conglomerate is about 90 feet thick in the sea cliffs at the entrance to Fossil Bay on Sucia Island (Figs. 69, 70). The conglomerate and overlying marine sandstone and shale dip to the NE toward the central axis of the Sucia syncline.

Pebbles in the conglomerate are almost entirely blue-gray phyllite, white quartz, and greenschist (Fig. 71) derived from Darrington phyllite and Shuksan greenschist, metamorphic rocks making up much of the North Cascades to the east. The average pebble size ranges from about one inch to several inches in diameter. Larger boulders are composed almost entirely of white, milky quartz, which stand out in strong contrast to the surrounding bluish-gray, smaller particles. The source of the pebbles in the conglomerate is obviously from underlying greenschist, phyllite, and quartz veins and pods in the phyllite.

55

Fossil Bay
sandstone

Sucia conglomerate

Figure 68. Sucia conglomerate and Fossil Bay sandstone in bluffs at the southwest end of the peninsula at the entrance to Fossil Bay, Sucia Island.

Figure 69. Sucia conglomerate at the entrance to Fossil Bay, Sucia Island.

Figure 70. Sucia conglomerate underlain by sandstone west of the entrance to Fossil Bay, Sucia Island. (Photo by George Mustoe)

Figure 71. Sucia conglomerate at Fossil Bay. G=greenschist, P=phyllite, Q=quartz. (Photo by George Mustoe)

The coarseness of the Sucia conglomerate means that the constituent pebbles and cobbles didn't travel very far from their source and must have been derived from underlying metamorphic Shuksan greenschist and Darrington phyllite. However, the conglomerate is underlain by sandstone and doesn't lie directly on the underlying metamorphic rocks.

Fossil Bay sandstone

The Fossil Bay sandstone lies directly on Sucia conglomerate at the entrance to Fossil bay. It consists of 365 feet of marine sandstone and shale (Fig. 72) interbedded with hard limestone beds 6-10 inches thick. Marine shell fossils are abundant in these rocks and can be easily seen along the south shoreline of Fossil Bay. Much of the shale has a high lime content and are tightly cemented.

Figure 72. A. Fossil Bay sandstone and limy shale. B. Fossils in Fossil Bay shale. Entrance to Fossil Bay, Sucia Island. (Photo by George Mustoe).

Fossils

Nanaimo sedimentary rocks and the fossils they contain have been studied for more than 150 years. Newberry (1857) first recognized Cretaceous fossils and McLellan (1927) mapped the Nanaimo rocks on Sucia, Waldron, Stuart, Orcas Islands and many smaller islands: Satellite, Johns, Ripple, Cactus, Flattop, Bare, Gull Rock, White Rocks, Skipjack, Parker Reef, and Sucia Islands.

Marine fossils are abundant in the Nanaimo Group marine sandstone at Fossil Bay on Sucia Island. They were first described in 1857, and many later fossil studies followed. The fossils consist of 48 species of clams (Pelecypods), 46 species of snails (Gastropods), 28 species of nautiloids (Cephalopods), 5 species of Brachiopods, 6 species of Crustacea, and one species of fish.

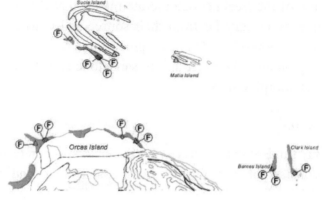

Figure 73. Fossil localities in Nanaimo Formation on
Orcas, Sucia, Clark, and Barnes Islands.

The sandstone and shale beds contain many marine fossil shells (Figs. 74-77).
Clam shells (*Inoceramus*) (Figs. 74, 75) are the most common fossils and are found
in abundance. *Baculites (*Fig. 76*)*, a straight shelled, nautiloid, and fossil oyster
(Fig. 77) are found throughout the beds. Also found, but considerably rarer, are
ammonites (Fig. 78, 79), a type of coiled nautiloid. All of these fossils indicate a
late Cretaceous age of the beds. Fossil wood and trees also occur in the Nanaimo.

Figure 74. *Inoceramus* fossils, Fossil Bay,
Sucia Island.

Figure 75. Fossil *Inoceramus*, Fossil
Bay, Sucia Island. (Photo by George

Figure 76. Baculites, a straight-shelled
nautiloid, Fossil Bay, Sucia Island.

Figure 77. Fossil oyster, Fossil Bay, Sucia.
Island. (Photo by George Mustoe)

Figure 78. Fossil ammonite, a coiled nautiloid, Fossil Bay, Sucia. (Photo by George Mustoe)

Figure 79 Typical coiled nautiloid.

Age

Abundant marine fossils from Orcas, Sucia, Stuart, Waldron, Clark, and Barnes Islands have been studied for more than a century and their ages are well known. All of the fossils consistently indicate a late Cretaceous age for the Nanaimo Formation.

The time of crystallization of zircon crystals in igneous rocks can be measured by isotope analysis or by counting tracks made by radioactive disintegration of uranium in zircon crystals (fission tracks). If the zircon is later eroded and deposited along with other sand grains, the time of deposition must be younger than the age of the earlier-crystallized zircon. Fission track dating of zircon in Nanaimo sandstone on Orcas Island indicates an age of 75.8 ± 5.6 million years and zircon in sandstone on Barnes Island were dated at 105 ± 10 million years.

CHUCKANUT FORMATION
Eocene
~50 million years

Fifty million years ago, the area that is now the San Juan Islands looked very different than it does today. The Cascade Range had not yet been formed and a broad alluvial plain covered the region in which the sediments were accumulating. Streams deposited sand, silt, coal, and pebbly gravel thousands of feet thick in a constantly subsiding basin. Palm trees grew amid forests of ferns and other tropical vegetation on the plain. Thick organic matter accumulated in large swamps that would become coal millions of years later.

No marine sediments occur in the Chuckanut sandstone so the rate at which sediment was delivered to the basin by streams was about equal to the rate of subsidence of the basin for millions of years until more than 10,000 feet of sand, gravel, and mud had been deposited. Groundwater percolated through the sediments over millions of years and precipitated chemicals as cement that bound the grains together to form hard rock. Burial under thousands of feet of overlying sediment compressed mud to form shale and the organic matter in swamps became coal.

Like the underlying Nanaimo sandstone, Chuckanut sandstone (Figs. 80, 81) is an arkosic sandstone, made up mostly of feldspar and quartz derived from erosion of nearby granite hills. Cross–bedding, which forms when streams deposit sand and gravel over the backside of a sand bar, is common in the sandstone (Figs. 82, 83).

Figure 80. Tilted Chuckanut sandstone at Echo Bay, Sucia Island.
(Photo by George Mustoe)

Figure 81. Steeply dipping Chuckanut sandstone and conglomerate, Ewing Island, Sucia.

Figure 82. Cross-bedding in Chuckanut sandstone on Sucia Island.
(Photo by George Mustoe)

Figure 83. Cross-bedding in Chuckanut sandstone, Sucia Island. Honeycomb
weathering of sandstone at the top of the beds. (Photo by George Mustoe)

Fossils in the Chuckanut Sandstone

Chuckanut sandstone is entirely non–marine—no marine fossils have been found anywhere in it. However, plant fossils are abundant. Leaf fossils are most common (Figs. 84–88), including palm fronds (Figs. 84-85), testifying to a climate more tropical than today's. However, because continents are known to drift for considerable distances, the palm trees may have been growing much farther south than where we see palm fossils today. Fossil logs (Fig. 86), some in growth position, occur in a few places, but are relatively rare. The fossil plants are typical of the early Tertiary period.

Figure 84. 50–million–year–old palm frond in Chuckanut sandstone.

Figure 85. Palm frond in Chuckanut sandstone (Photo by George Mustoe)

Figure 86. Fossil logs in Chuckanut sandstone, Sucia Island. (Photo by George Mustoe)

64

Figure 87. Plant fossils in Chuckanut sandstone. (Photo by George Mustoe)

Age

Zircon in Chuckanut sandstone on Sucia Island has been dated at 58.0 ± 6.0 million years. Four fission-track ages from Chuckanut sandstone on the mainland south of Bellingham ranged in age from 50 to 55 million years. All of these dates confirm the Eocene age of the Chuckanut Formation.

THE ICE AGE IN THE SAN JUAN ISLANDS
Pleistocene (100,000 to 10,000 years ago)

During the Pleistocene Ice Age, which began about 2.5 million years ago, global temperatures dropped dramatically and immense continental ice sheets more than 10,000 feet thick covered vast areas of northern North America (Fig. 88), Europe, and Eurasia. Accumulations of ice and snow in British Columbia built the Cordilleran Ice Sheet that advanced southward into Washington at least six times (and perhaps more), only to melt away thousands of years later, leaving as their only footprint scouring of the land surface and glacial debris deposited by the ice. Sediments of each successive glaciation buried previous ones so that only deposits of the last glaciation are exposed on the land surface, and evidence of the older glaciations may be seen only in sea cliff exposures.

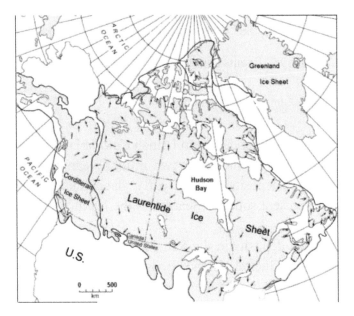

Figure 88. Huge continental glaciers of the last Ice Age in North America.

The enormous size of these ice sheets dwarfs the small alpine glaciers in the Cascade Range. The Cordilleran Ice Sheet, which filled the Puget Lowland from the Olympic Mts. to the Cascades and from the Canadian boundary to Olympia, was only a lobe of a still larger ice sheet in British Columbia (Fig. 88). On a hot summer day, it is difficult indeed to imagine that ice more than a mile thick once covered the San Juan Islands. Yet deposits of rock debris left by the ice give indisputable testimony that such immense glaciers occupied the lowland not only once, but at least six times and perhaps more. Periods of glaciation were separated by interglacial warm periods, similar to or warmer than the present (Fig. 89).

FRASER GLACIATION	**SUMAS STADE**	Sumas glacial deposits ~11,400 to ~10,000 ^{14}C years ago • Emergence of the northern Puget Lowland above sea level. • Readvance of the Cordilleran Ice Sheet and deposition multiple moraines in Whatcom County.
	EVERSON INTERSTADE	• Shorelines cut at multiple levels now 300' above sea level 11,500 to 11,700 ^{14}C years ago Everson glaciomarine deposits 11,700 to ~12,500 ^{14}C years ago • ~400' of submergence of the San Juan Islands and deposition of Everson glaciomarine sediments from floating ice
	VASHON STADE	Vashon glacial deposits ~17,500 to ~12,500 ^{14}C years ago • Advance of the Cordilleran Ice Sheet, 6000' thick • ~400' of depression of the land due to weight of the ice sheet • Deposition of glacial sediments from the ice sheet • Scouring and grooving of bedrock by the glacier

Deposition of silt and sand by streams and lakes

Figure 89. Events and deposits of the last Ice Age.

Older glaciations

Continental ice sheets extended from British Columbia into the Puget Lowland south of Seattle for the first time about two million years ago, followed by five subsequence advances. Deposits of these early glaciations are absent in the San Juan Islands but are present south of Seattle, so the ice must have passed over the islands.

The oldest Ice Age deposits in the San Juan Islands are exposed in sea cliffs along the west coast of Guemes Island. Beds of interglacial silt and sand of the Whidbey Formation are exposed at the base of sea cliffs at Yellow Bluff (Figs. 90–92). A limiting age of older than 40,000 years was obtained from a three-foot thick peat bed at Yellow Bluff (Figs. 93, 94). A more accurate age of 105,000 years was measured in the Whidbey Formation at the south end of Whidbey Island.

About 100 feet of sand and gravel deposited by glacial meltwater streams overlies the Whidbey interglacial sediments at Yellow Bluff (Figs. 90–92). The age of the sand and gravel has not been measured there, so their age is not known for certain. The sand and gravel is most likely from the Possession Glaciation (70–90,000 years ago).

During the nonglacial interval between the Possession and Fraser Glaciations, known as the Olympia nonglacial interval, floodplain sand, silt, clay, and peat were deposited in the San Juan Islands. A radiocarbon date of 22,940 years was obtained from wood fragments in these sediments at Indian Cove on Shaw Island.

Figure 90. Interglacial silt, sand, and peat of the Whidbey Fm. (gray) overlain by outwash sand and gravel (brown), Yellow Bluff, Guemes Island.

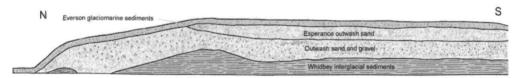

Figure 91. Geologic cross section at Yellow Bluff.

Figure 92. Ice Age sediments making up Yellow Bluff, Guemes Island. The Whidbey Fm. stream sand and silt is about 100,000 years old. Gmd=glaciomarine pebbly silt 14,000 years old.

Figure 93. Whidbey Fm. silt and sand overlain by peat. 70–90,000–year–old
outwash sand and gravel at top, Yellow Bluff, Guemes Island.

Figure 94. Peat bed at the top of the Whidbey Fm., Yellow Bluff.

Fraser Glaciation–The last advance and retreat of the Cordilleran Ice Sheet 20,000 to 11,500 years ago

The last major phase of the Ice Age in San Juan Islands occurred during the Fraser Glaciation, which began with growth of the Cordilleran Ice Sheet in British Columbia about 20,000 years ago and ended 11,500 years ago. Prior to that, ~20,000 to ~30,000 years ago, the Puget Lowland, including the San Juan Islands, was ice-free and stream, marine, and lake deposition and erosion operated much as they do today. As the ice sheet grew, it advanced southward across the US-Canadian border into Washington and at its maximum, extended just south of Olympia (Fig. 95), filling the entire Puget Lowland from the Olympic Mts. to the Cascades.

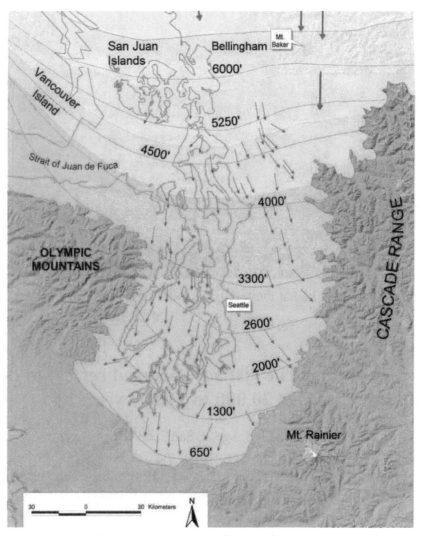

Figure 95. The Cordilleran Ice Sheet in northwest Washington. Blue area is glacial ice; numbers are elevations of the ice surface above sea level.

VASHON MAXIMUM OF THE CORDILLERAN ICE SHEET

During the Vashon glacial advance, the Cordilleran Ice Sheet extended from British Columbia into Washington and reached its maximum extent south of Olympia, about 140 miles south of the Canadian border (Fig. 95). At that time, only mountain peaks above 6,000 feet in the North Cascades stood above the surface of the ice, as indicated by erratic boulders now strewn on their slopes. The Vashon glacier passed over the San Juan Islands sometime after 20,000 years ago and reached its maximum size about 17,000 years ago. As the ice thickened and moved southward, it filled the Puget Lowland with ice more than a mile thick and inundated large areas of the North Cascades near Mt. Baker (Fig. 95). The glacier buried the San Juan Islands under at least 6000' of ice (Fig. 96).

71

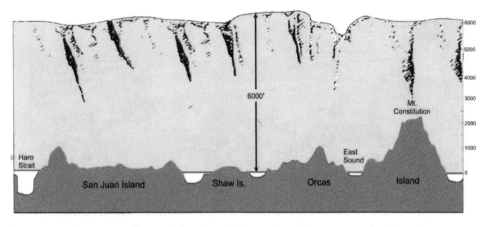

Figure 96. Cross section of the Cordilleran Ice Sheet over the San Juan Islands. Ice was 6000' thick as it rode over the islands 17,000 years ago.

Ice erosion

As the ice flowed over the San Juan Islands, it scoured and polished the bedrock and carved many grooves and striations, which persist today on rock exposures near the coastline (Figs. 97, 98). The grooves and striations made by the ice sheet record the direction of ice movement at the time they were made. Elongate hills were also sculpted by flowing ice and show a north–south direction of ice movement.

Figure 97. Ice–polished, grooved bedrock, Cattle Point, San Juan Island. Arrow shows direction of ice movement

Figure 98. Glacial groove carved in bedrock by the Cordilleran Ice
Sheet, Cattle Point, San Juan Island.

Deposition of glacial sediments

Large amounts of rock debris eroded in British Columbia were transported
southward by the Cordilleran Ice Sheet and deposited in the San Juan Islands. Large
individual boulders, known as erratics, were strewn throughout the islands and are
now recognizable as such by the composition of the rocks, which are not native to
the islands (Figs. 99, 100). The largest known erratic, about 65 feet wide and 63
feet tall, rests on the sea floor south of Lopez Island.

Figure 99. Granite erratic boulder from British Columbia, Fossil Bay, Sucia Island.

Figure 100. Large erratic boulder, Decatur Island.

Figure 182. Granite erratics strewn over the landscape, Iceberg Point, Lopez Island.

Rock debris carried by the Cordilleran Ice Sheet was deposited as a blanket of glacial till and glaciomarine sediments over much of the islands (Fig. 102). These may be recognized by their concrete-like appearance, with pebbles and cobbles imbedded in a matrix of clay, silt and sand (Fig. 103). The glacial till is well compacted as a result of the weight of thousands of feet of ice pressing on it, but the glaciomarine drift was deposited from floating ice so is not compact. Rock debris carried by the ice was ground up by crushing of rocks by the ice. Many of the rocks carried by the ice were ground down to flat surfaces (faceted), polished, scoured, and scratched.

GEOLOGIC MAP
OF THE
SAN JUAN ISLANDS

Figure 102. Portions of the San Juan Islands mantled with glacial deposits (yellow) (Modified from DNR map, 1975).

Figure 103. Glacial till deposited by the Cordilleran Ice Sheet about 15,000–20,000 years ago. The till has a concrete like appearance and is compact. The composition of pebbles and cobbles in the till indicate that they came from British Columbia.

As the ice rode over the sand and gravel, the force of the moving ice deformed the sediments beneath it (Fig. 104).

Figure 104. Deformation of sand and silt overridden by the Cordilleran Ice Sheet. Beds at the base of the bluff have been tilted from horizontal to vertical and the overlying beds have been folded and shoved over the vertical beds beneath. The beds just above the contact with the vertical beds are now upside down.

RAPID RETREAT AND DISINTEGRATION
OF THE CORDILLERAN ICE SHEET

About 15,000 years ago, the massive Cordilleran Ice Sheet began to feel the effects of abrupt and intense climatic warming and underwent rapid, large–scale melting and recession. The ice sheet melted back from its terminus near Olympia and receded past the Seattle area. As the terminus of the glacier retreated, the ice also thinned rapidly. Relative sea level at that time was several hundred feet higher than at present, and as the ice retreated to the vicinity of Whidbey Island, it thinned enough that marine water entered the lowland from the Strait of Juan de Fuca and floated the ice, causing wholesale, rapid disintegration of the remaining ice sheet (Fig. 105). Rapid disintegration of the Cordilleran Ice Sheet occurred in deep water all the way to Canada and the region was covered with floating ice (Fig. 105). Ice remained locally grounded in some of the eastern part of the lowland and San Juan Islands. The disintegration of the Cordilleran Ice Sheet was probably similar to modern breakups of tidewater glaciers in Alaska. When glaciers reach a critical thickness, allowing marine water to extend under the terminus, large masses of the lower glacier break off and float away from the terminus. This has happened many times to Alaskan glaciers, which serve as a modern analog to the breakup of the Cordilleran Ice Sheet.

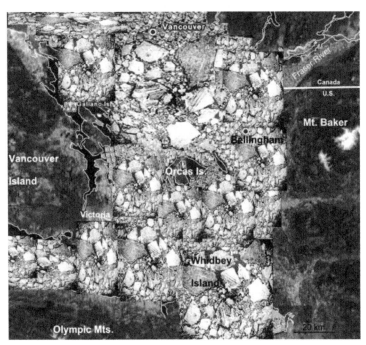

Figure 105. Reconstruction of the disintegration of the Cordilleran Ice Sheet by floating of the ice in marine water about 13,000 years ago. The white areas represent disintegrating floating ice in sea levels higher than today. When the floating ice melts, rock debris is deposited on the sea floor as glaciomarine stony clay.

EVERSON GLACIOMARINE DEPOSITS

Submergence of the San Juan Islands beneath the sea 12,500 years ago

When the Cordilleran Ice Sheet thinned, floated, and disintegrated, floating ice deposited poorly sorted sediments on the sea floor (Fig. 106), burying mollusc shells in stony clay. Debris released by melting of the ice sank to the sea floor below, depositing a heterogeneous mixture of clay, sand, pebbles, cobbles, and boulders resembling glacial till (Figs. 107–109). We know that this sediment was glacial because many of the stones in the glaciomarine sediments have been glacially faceted and polished (Fig. 111) and that the source was British Columbia because the rock types could only have come from the north. We also know that the sediments were deposited in marine water because marine fossils (Fig. 110) occur throughout the deposits.

Thicknesses of the glaciomarine sediments reach 25 feet or more on some of the islands. Figure 106 shows the depositional environment of the sediments. At the highest sea level stand during the Everson glaciomarine interval, sea level was about 400 feet above present sea level and all of the islands below that were submerged beneath the sea.

The age of the Everson glaciomarine stony clay is well established at 11,700 radiocarbon years (about 14,000 calendar years) by more than 40 radiocarbon dates on wood and shells in the deposits.

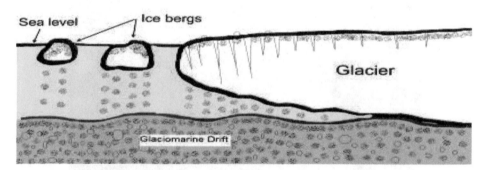

Figure 106. Depositional environment of glaciomarine sediments. As floating ice melted, clay, sand, pebbles, and cobbles in the ice were dumped on the sea floor, burying marine organisms on the bottom.

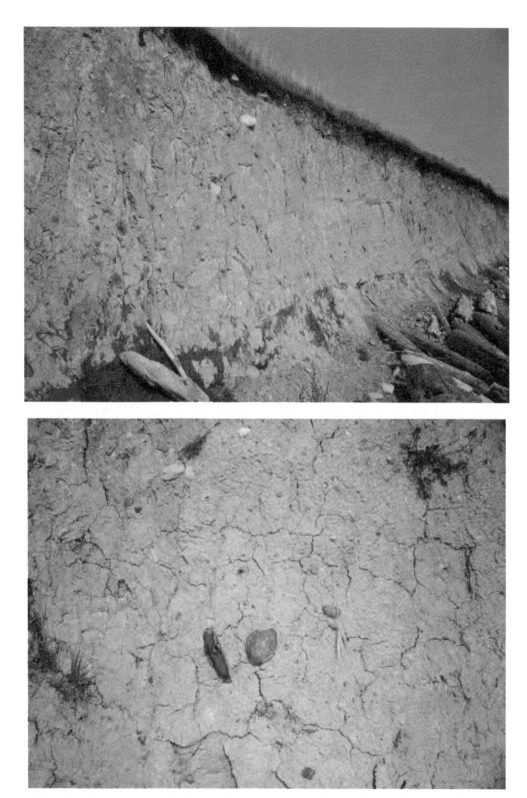

Figure 107. Everson glaciomarine stony clay, Cattle Point, San Juan Island.

Figure 108. Everson glaciomarine poorly sorted silt, sand, pebbles cobbles, and boulders containing many clam shells..

Figure 109. Everson glaciomarine sediments containing many marine fossils.

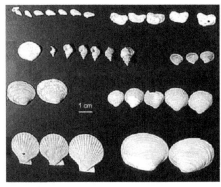

Figure 110. Marine shell fossils from glaciomarine sediments.

Figure 111. Glacially abraded pebbles from Everson glaciomarine sediments.

OLD SHORELINES–FOOTPRINTS OF ANCIENT SEA LEVEL

The Everson glaciomarine interval is well dated at 11,700 radiocarbon years ago. The marine submergence during the Everson in the San Juan Islands came to a close as the land rose when the weight of thousands of feet of glacial ice disappeared. During the height of the glaciation of the last Ice Age, the weight of the ice sheets caused the land to sink hundreds of feet. When the ice melted away, rebound of the land rose to former levels. Thus, when the Cordilleran Ice Sheet melted away the San Juan Islands began to rebound, leaving old shorelines at elevations up to 300 feet above present sea level and many successively lower shorelines that record the emergence of the islands from the sea. The San Juan Islands at that time looked very different than they do now (Fig. 112)

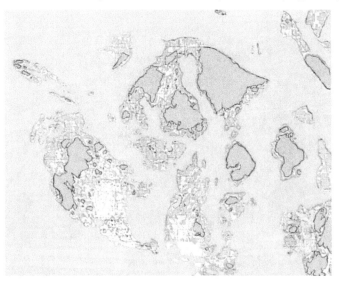

Figure 112. Reconstruction of what the San Juan Islands would have looked like about 14,000 years ago. Most of the islands were submerged 300 feet and only the higher hills stood above sea level (brown areas).

Laser-generated images of the San Juan Islands show multiple raised shorelines. They are best developed at Cattle Point on San Juan Island (Fig. 113), where the highest shorelines extend to elevations of 300 feet. Another dozen shorelines occur below that. Similar shorelines also occur near Roche Harbor, Friday Harbor, and False Bay on San Juan Island (Figs. 114–126).

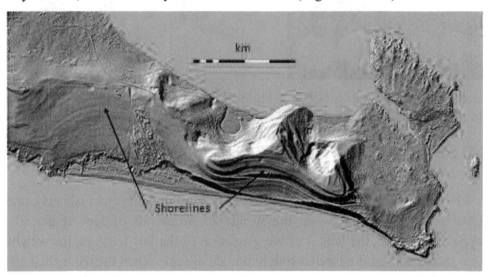

Figure 113. LIDAR image of multiple raised shorelines at Cattle Point on San Juan Island. The highest shoreline extends to an elevation of 300 feet.

Figure 114. Multiple raised marine shorelines at Cattle Point, San Juan Island.

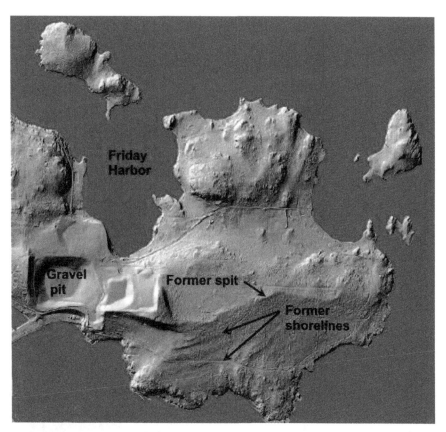

Figure 115. Raised shorelines, Friday Harbor, San Juan Island.

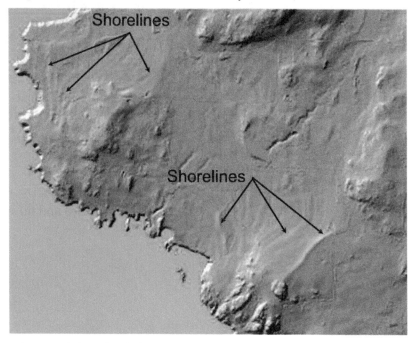

Figure 116. Raised shorelines, near False Bay, San Juan Island.

Well-developed raised marine shorelines are present along the northern coast of Orcas Island. The highest shorelines occur at 300 feet between Point Doughty and East Sound (Fig. 117). Numerous shorelines show up as low concentric ridges of ancient beach sand and gravel resembling rings in a bathtub.

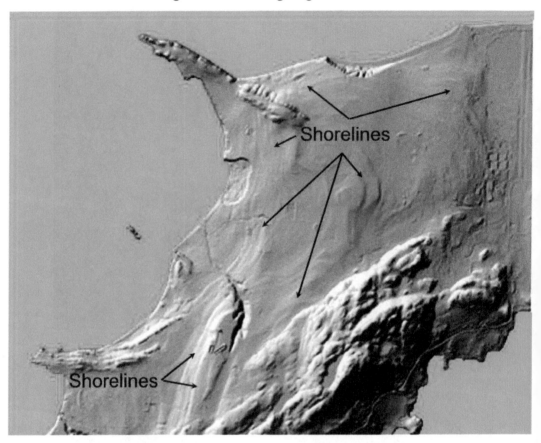

Figure 117. LIDAR image of multiple raised shorelines near Point Doughty on northwestern Orcas Island. The highest shorelines extend to elevations of 300 feet above present sea level. Their age is 11,500 to 11,700 ^{14}C years before present.

Distinct shorelines are also visible on Lopez (Figs. 118–121), Decatur (Fig. 122), northern Blakely (Fig. 123), Lummi Island (Figs. 124–126) and other islands.

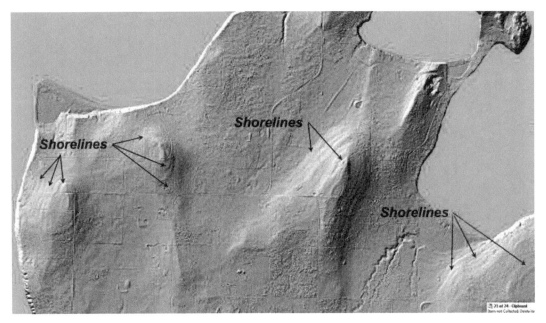

Figure 118. Raised shorelines on northern Lopez Island.

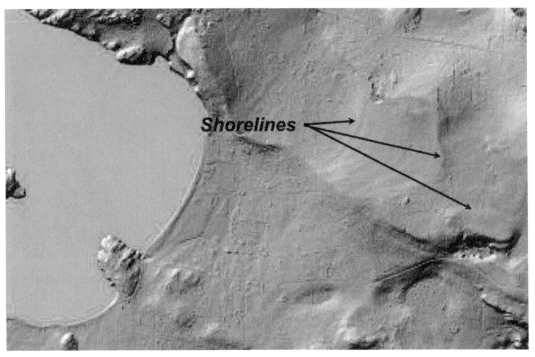

Figure 119. Raised shorelines, southern Lopez Island.

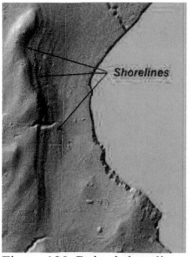

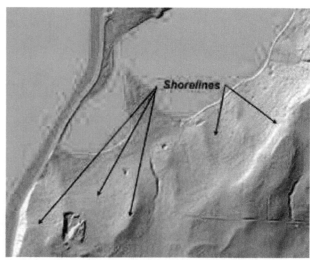

Figure 120. Raised shorelines, east–central Lopez, Island.

Figure 121. Raised shorelines, Fisherman Bay, Lopez Island.

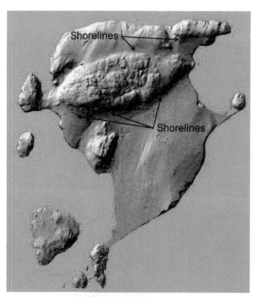

Figure 122. Raised shorelines, Decatur Island.

Figure 123. Raised shorelines, Blakely Island. The highest shoreline is 300 ft.

Figure 124. Raised shorelines, northern Lummi Island.

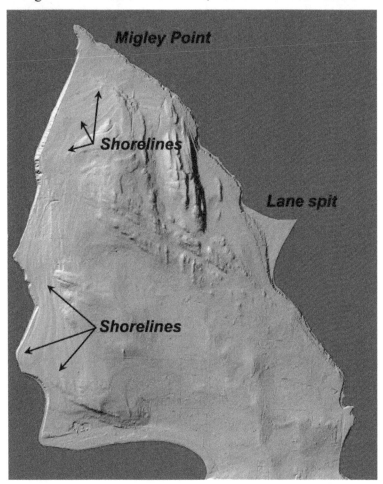

Figure 125. Raised shorelines of northern Lummi Island shown on lidar image

EARLY INHABITANTS

Soon after the Cordilleran Ice Sheet had melted away, sea level in the San Juan Islands stood 300 feet higher than at present and shorelines were formed on virtually all of the higher islands (see the section on old shorelines above). Because of the high relative sea level, the San Juan Islands looked considerably different than they do now (Fig. 112). 14,000 years ago, the Puget Lowland, including the San Juan Islands, was inhabited by numerous large mammals after the Cordilleran Ice Sheet melted away. Bones and skeletens of wooly mammoths, bison, giant ground sloths, and large bears have been recovered from Ice Age deposits in many places.

Bones and skeletal remains of extinct bison and giant sloths have been found in peat bogs at five localities on Orcas Island (Fig. 127).

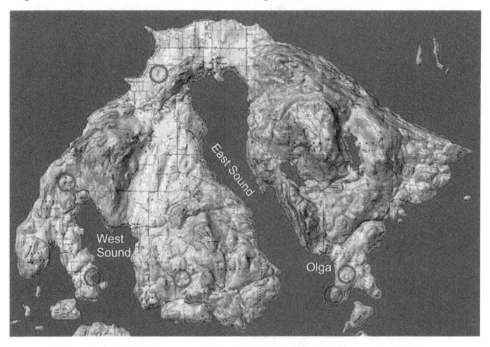

Figure 127. Localities where bones of extinct bison and sloths have been found.

Ayers Pond

While excavating a peat bog to make a pond, bison bones were discovered at Ayers pond near Olga on Orcas Island (Figs. 128–130). The curious thing about the bones was that some showed signs of fracturing and chipping that have been interpreted as evidence of human butchering. This means that humans occupied the islands at least 14,000 years ago. The skull and other bones are on display at the East Sound museum.

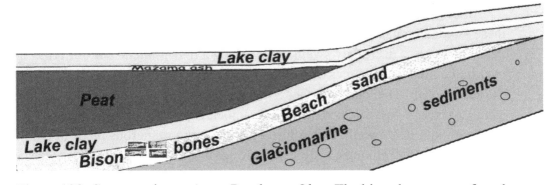

Figure 128. Cross section at Ayers Pond near Olga. The bison bones were found at the base of the peat bog resting on beach sand containing marine shells dated at 11,700 ^{14}C years (14,000 calendar years).

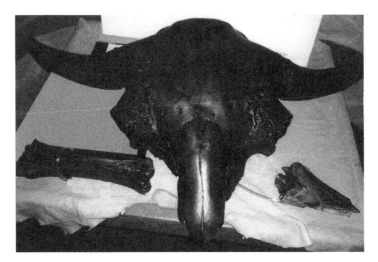

Figure 129. Bison skull from the base of a peat bog at Ayers pond, Orcas Island. (East Sound museum)

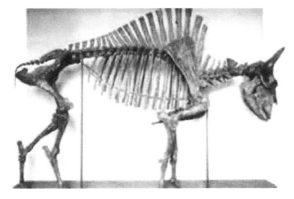

Figure 130. Skeleton of North American bison.

Among the interesting questions posed by these bone discoveries is how did these large mammals get to the islalnds? The most plausible answer appears to be that, like modern deer, they swam. Bison are good swimmers and can cross rivers over half a mile wide. Bones of a giant sloth were found in a peat bog southwest of Turtleback Mt. The peat overlies Everson glaciomarine sediments containing abundant marine fossils dated at 11,700 radiocarbon years.

Mammoth tusks, teeth, and bones are common in numerous place in the Puget Lowland, especially along coastlines where sea cliffs of glacial sediments are being eroded by waves. Woolly mammoths, bison, giant sloths, sabre toothed tigers, and a number of other large mammals died out at the end of the last Ice Age as part of a mass extinction of large mammals in northern latitudes. Although rapidly changing environments and human hunting have been suggested as possible causes of this mass extinction, the cause remains unknown.

HOLOCENE EPOCH

SHORELINE FEATURES

Relentless waves breaking against the shorelines of the San Juan Islands for many thousands of years have sculpted the coasts of the islands. Wave energy expended at the base of sea cliffs undercuts slopes along the shore forms sea cliffs that retreat landward under continued wave attack. Sand and gravel are dragged back and forth along beaches, pushing it up the beach with each wave and pulling it seaward again as the water runs back. The net effect is analogous to a great saw, held horizontally, the teeth of which are the loose particles moved by the waves. Prolonged wave erosion causes retreat of sea cliffs and creates wave-cut benches, beaches, spits, and bars along the coastline.

COASTAL EROSION

The masses of water tossed against the land by the incessant pounding of waves against shorelines expend significant amounts of energy that effect coastline morphology as a result of the mechanical erosion that it creates. Coastal erosion is most pronounced during storms when wave energy is high and previously weathered material is removed and newly exposed material is eroded. Storm waves are especially effective in eroding unconsolidated sediments and highly fractured or bedded rocks which are vulnerable to breaking loose as the impact of waves generates high water pressure.

Sea Cliffs

As wave erosion undercuts slopes at shorelines, sea cliffs are developed that progressively retreat landward. Waves commonly erode an overhanging, wave–cut notch in rocks (Fig. 131), leaving a permanent mark that provides evidence of former sea levels. Undercutting of the base of sea cliffs by waves promotes accelerated down-slope movement of loose rock debris, which collects at the base of bluffs until removed by wave erosion.

The rate of sea cliff recession is controlled primarily by the vigor of wave action and the resistance of the cliff–making material. Sea cliff erosion rates are quite variable. Bluffs composed of very resistant material under weak wave activity retreat very slowly, whereas shorelines composed of less resistant, unconsolidated material being eroded by vigorous wave action retreat very rapidly.

Figure 131. Wave–cut notch in Nanaimo sandstone, Sucia Island.

Generally, crystalline rocks (granite, volcanic rocks) show the lowest rates of sea cliff retreat, and unconsolidated Quaternary sediments show the highest rates. Most bedrock sea cliffs show little retreat during a human lifetime, but cliff recession in unconsolidated sediments average about a foot per year, although the recession is highly variable over short periods of time.

Wave cut benches

As sea cliffs retreat under wave erosion, they leave behind wave–cut benches beveling rocks on the sea floor slightly below high tide level (Figs. 132 and 133). As waves break on the shore, sand and pebbles on the sea floor are dragged back and forth with each passing wave, abrading the rock like a great horizontal saw, the teeth of which are the loose grains moved by the waves.

Figure 132. Receding sea cliff and erosion of a wave–cut bench, Fossil Bay, Sucia Island.

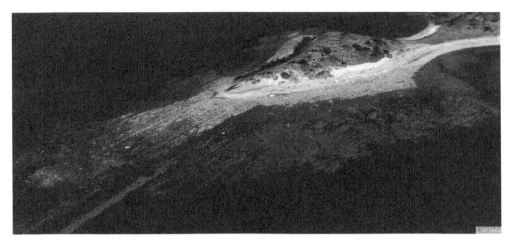

Figure 133. Extensive wave-cut bench, Little Sucia Island.

COASTAL DEPOSITION

BEACHES

Beaches are accumulations of sand, pebbles, and/or cobbles along a shoreline. They consist of whatever sediment is available for movement by wave action. Some are built from material derived from erosion of nearby sea cliffs, while others are made of sediments carried by streams to the coast (Fig. 134, 135).

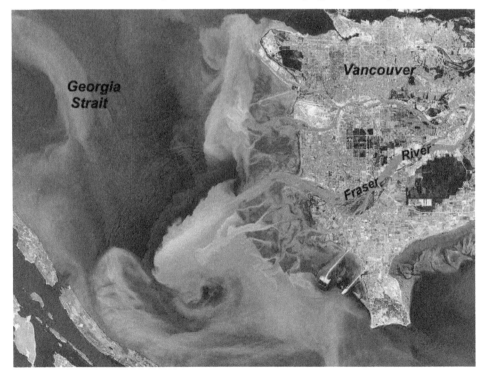

Figure 134. Sediment plume from the Fraser River. (NASA)

Figure 135. Double spits in Bellingham Bay extending from Point Francis (to the left). A plume of muddy water to the right of the outer spit is from floodwaters of the Nooksack River.

The location of beaches is determined by sediment supply and wave activity. Beaches are dynamic features, frequently changing to adapt to changing conditions. The loose sediment composing them is mobile and subject to movement, so beaches represent an equilibrium between wave action and sediment supply.

LONGSHORE DRIFTING

Waves are generated by wind, and because of the great variety in the angle that shorelines make with the wind, most of the time waves approach a coastline at some oblique angle. The effects of this oblique wave angle on coasts are profound. As waves break on a beach at an oblique angle they push up the beach at an angle to the waterline. But as the water from each wave runs back down the beach toward the waterline, the water moves directly down the beach face in the direction of the beach slope, rather than retracing the oblique path the water took as the wave swashed up the beach face (Fig. 136). The net effect is that particles end up several inches down the beach from where they started. These relatively small increments of lateral movement by each wave are then multiplied by the almost infinite repetitions brought about by each wave that swashes up the beach face and the end result is the movement of enormous quantities of sediment laterally along the shoreline. The amount of sediment moved by longshore drifting may be quite large. Along the east coast of the United States, an estimated 450,000 cubic yards of sediment is moved by longshore drift every year. That's equivalent to 41,000 dump trucks of sediment every day and 150,000 dump trucks every year. Because many of the shorelines in the San Juan Islands are rocky, the amount of sediment moved laterally along the coast by waves is much less, but the process is the same and results in deposition of spits and bars (Fig. 137–140).

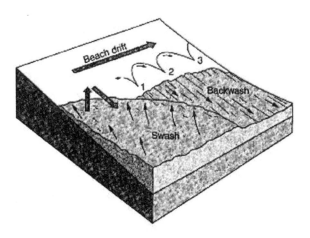

Figure 136. Longshore drifting. A wave washes up on the beach face at an oblique angle, pushing sediment obliquely up the beach face (left-most arrow), then washing back down the beach at right angles to the slope, dragging sediment with it (second arrow from the left), resulting in net transport laterally along the beach.

SPITS AND BARS

Where shorelines make sharp bends, longshore transport of sediment continues to move in the same direction instead of continuing around the corner and is deposited in the deeper water offshore. As the sediment builds seaward a *spit* forms, connected at one end to land and terminating in open water at the other end.

Figure 137. Spit extending from Point Frances, Portage Island, Bellingham Bay.

Figure 138. Spit at Fisherman Bay, Lopez Island.

Figure 139. Flat Point, Lopez Island, made by converging directions of longshore transport of sediment to build a spit seaward. Note the lagoon and bog enclosed by the spit.

Figure 140. Lane spit extending seaward from Lummi Island as a result of converging directions of longshore transport of sediment.

Tombolos

Once formed, a spit continues to extend in the direction of longshore drifting. Some spits grow from the land toward offshore islands, eventually connecting the islands to the land as *tombolos*. The reaching out of spits to attach onto islands is not random but results from the bending of waves around islands that affects the directions of longshore sediment transport on the beach (Fig. 141). As waves approach the shore, they encounter shallower water in the vicinity of islands and are slowed down, which causes them to bend around the island (Fig. 141) and results in longshore drifting toward the lee of the island. Because the direction of longshore transport on both sides of the beach is toward the lee of the island, sediment moving laterally along the beach accumulates there, conforming to the shape of the wave pattern. As more and more sediment is deposited, the spit extends seaward toward the island (Figure 141–145) until it eventually reaches it and attaches the island to the mainland as a *tombolo*.

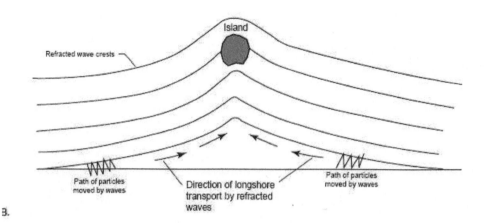

B.

Figure 141. Effect of an offshore island on the wave pattern. Waves are bent as they encounter the island, causing waves to drive sediment in converging directions in the lee of the island.

Figure 142. Spencer spit reaching out to Frost Island from Lopez Island.

Figure 143 Tombolo, Decatur Island.

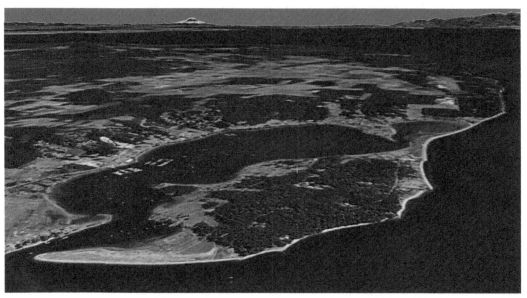

Figure 144. Tombolo at Fisherman Bay, Lopez Island.

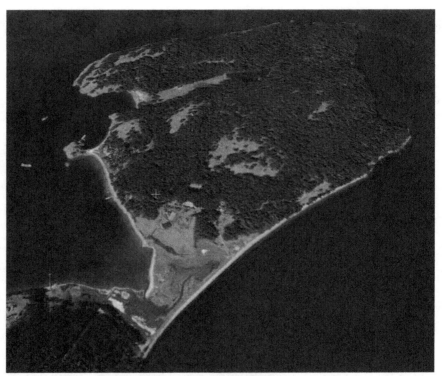

Figure 145. Tombolo at Humphrey Head, Lopez Island.

GEOLOGIC FRAMEWORK
OF THE SAN JUAN ISLANDS

This section contains much new, previously unpublished, LIDAR, sonar, and satellite data. It is aimed primarily at geologists and may be a bit difficult for non-geologists to follow. In that case, you may want to skip through this section (noting the faults) and move on to the rest of the book dealing with the geology of individual islands.

The rocks of the San Juan Islands have been intensely deformed—folded, faulted, and internally sheared. The oldest rocks are, in general, the most deformed. Rocks of the Turtleback Crystalline Complex, East Sound Group, Garrison schist, Orcas Chert, Constitution Formation, and Carter Point–Lummi Formation have been intensely sheared and slightly metamorphosed under low heat and high pressure. In contrast, the Haro, Spieden, Deer Point, Nanaimo, and Chuckanut Formations have not been metamorphosed, internally sheared, or as intensely deformed.

Since 1975, the geological framework of the San Juan Islands has been thought to be the result of five major, extensive, low–angle thrust faults known collectively as the San Juan Thrust (Nappe) System. However, recent LIDAR and sonar data (detailed below) show that this thrust system does not actually exist.

101

THE SAN JUAN ISLANDS THRUST SYSTEM: NEW PERSPECTIVES FROM LIDAR AND SONAR IMAGERY

The concept of multiple thrust sheets stacked successively one upon another and bringing terranes from sources hundreds of miles away and shuffling them together like a deck of cards was first published in 1975 by Joe Vance and John Whetten. Five thrust sheets were mapped as discrete, single–plane, named thrust faults: the Rosario, Orcas, Haro, Lopez, and Buck Bay thrusts that were thought to extend over a wide area in the islands (Fig. 146). The source area of these thrust faults was postulated to be the North Cascade Range about 100 miles to the east.

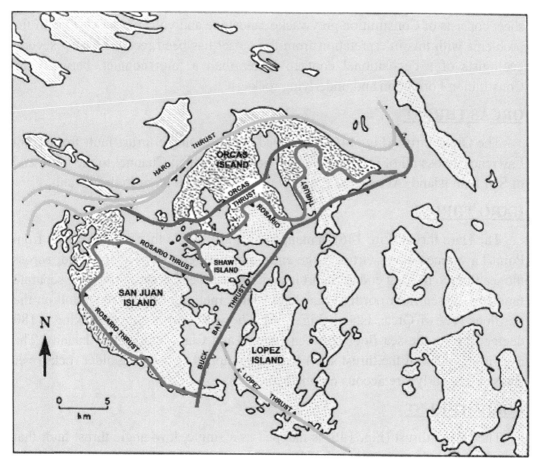

Figure 146. Postulated five thrust sheets in the San Juan Thrust (nappe) System. (Modified from Brandon et al., 1988)

POSTULATED THRUST FAULTS OF THE SAN JUAN THRUST (NAPPE) SYSTEM

The postulated five thrust faults were thought to have shoved large crustal slabs westward from the North Cascades during a period of crustal activity in the late Cretaceous. The thrust faults were thought to have brought far–distant rocks to the San Juan Islands.

ROSARIO THRUST

The Rosario thrust (Fig. 146) is mapped as a single, low–angle thrust fault that extends all the way across the San Juan Islands from Point Lawrence on Orcas Island to the southeastern tip of San Juan Island at Cattle Point. The proposed thrust fault follows a very loopy path across the islands, presumably as a result of erosion of a gently dipping fault plane. For the most part, the overriding block of the thrust sheet consists of Constitution greywacke sandstone and volcanic rocks. One of the problems with this interpretation from the outset has been recognition by several geologists of a depositional contact, rather than a fault contact, between the Constitution Formation and underlying rocks.

ORCAS THRUST

The Orcas thrust (Fig. 146) is mapped as a low–angle thrust fault from Point Lawrence across all of Orcas Island and across San Juan Channel to the north end of San Juan Island. At its eastern end, it merges with the Rosario thrust fault.

HARO THRUST

The Haro thrust (Fig. 146) is mapped as a low–angle thrust that extends from Point Lawerence along virtually the entire northern coast of Orcas Island, across the sea floor to the east end of Stuart Island. The west end is shown as three separate faults on the sea floor north of San Juan Island, merging into a single fault on the sea floor NW of Orcas Island (Fig. 146). The west end is shown making a 180 degree turn on the sea floor between Lummi and Clark and Barnes Islands. The overriding block of the thrust sheet has been mapped as bringing older rocks over Jurassic and early Cretaceous marine beds.

LOPEZ THRUST

The Lopez thrust (Fig. 146) is mapped as a single, low–angle thrust fault that extends across the southern end of Lopez Island, bringing Fidalgo ophiolite rocks over Jurassic and Cretaceous greywacke sandstone and pillow basalt. To the south, many parallel faults make high, conspicuous fault scarps but are also shown as thrust faults.

BUCK BAY THRUST

The Buck Bay thrust (Fig. 146) is mapped as a single, low–angle thrust from Cattle Point on San Juan Island along the entire length of Upright Channel, past Point Lawrence, ending between Lummi and Clark and Barnes Islands.

NEW DATA FROM LIDAR IMAGERY

A new form of making images of the Earth's surface has recently revolutionized geology. Laser beams, transmitted downward from an airplane, are recorded with special equipment that accurately measures the distance from the plane to the ground surface. These data can then be made into photo–like images by computers. The process, known as LIDAR, reveals remarkable topographic details not visible by any other means. LIDAR doesn't 'see' vegetation, so the resulting image is as though the land has been stripped totally bare, clearly revealing faults and other geologic structures. Faults show up exceptionally well on LIDAR imagery as linear gashes in the topography made by etching out of weak rock in fault zones and as long, straight scarps that cut across geologic structures. The straightness of the faults cutting across areas of topographic relief indicates that they dip steeply.

All of the geologic maps that have been made over the past forty years show multiple low–angle thrust faults, but very few of them show high–angle faults. This is partly due to the difficulty of recognizing them with the heavy vegetative cover of most of the islands. However, LIDAR does not 'see' the vegetative cover and shows that the San Juan Islands are riddled with numerous high–angle faults never before recognized.

ORCAS ISLAND FAULTS

LIDAR images show a number of long, linear, high–angle faults cutting across Orcas Island (Fig. 147). Very few of these faults have been previously mapped.

CENTRAL ORCAS FAULT

The Central Orcas fault (Fig. 147) makes a conspicuous continuous, straight, gash in the topography that extends from Point Lawrence across the entire width of the island to Deer Harbor. The straightness of the fault across rugged topography indicates that it dips at a high angle.

OLGA FAULT

The Olga fault (Fig. 147) is roughly parallel the Central Orcas fault and extends from Doe Bay on the east coast of the island to the submarine scarp that makes the south margin of Crane Island south of Deer Harbor. The straightness of this fault trace indicates that it is steeply inclined.

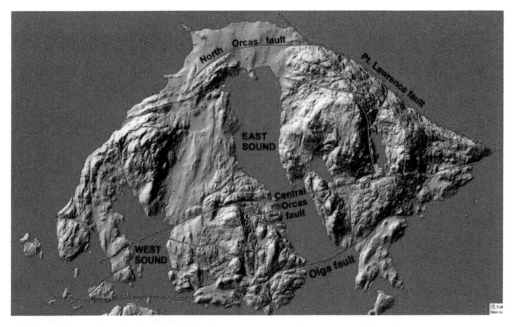

Figure 147. LIDAR image of Orcas Island showing major faults.

NORTHERN ORCAS FAULT

Nanaimo marine beds along the north shore of Orcas Island trend generally NW–SE and dip to the southwest, but their contact with older Paleozoic beds to the south is covered by Pleistocene glacial deposits. However, because the Nanaimo beds dip *into* the older rocks (Fig. 148), a fault apparently crosses the northern part of the island, separating the Nanaimo beds from the Paleozoic rocks to the south. This fault has been shown on previous geologic maps as the Haro thrust, but it is not exposed and thrust faulting in the Nanaimo beds is unknown elsewhere. Thus, the Northern Orcas fault appears to be a high–angle fault.

Figure 148. The northern Orcas fault.

SAN JUAN ISLAND FAULTS

Many long, high–angle faults that criss–cross San Juan Island are apparent on LIDAR imagery (Figs. 150, 151). These faults have not been mapped previously.

BEAVERTON VALLEY FAULT

The Beaverton Valley fault (Figs. 150, 151) extends E–W across the entire width of San Juan Island from Friday Harbor to Lime Kiln Point. The straightness of the fault across rugged topography indicates that it is steeply dipping.

FRIDAY HARBOR FAULT

The Friday Harbor fault (Figs. 150, 151) is roughly parallel to the Beaverton Valley fault and extends E–W across most of San Juan Island. It too remains straight across topographic relief, indicating that it is steeply dipping.

SPORTSMAN LAKE FAULT

The Sportsman Lake fault (Figs. 150, 151) is parallel to the Beaverton Valley and Friday Harbor faults and extends the width of San Juan Island. It too is a steeply dipping fault.

Numerous other smaller, high–angle faults that criss–cross the island, including several N–S faults, are also apparent on LIDAR images Figs. (149, 150).

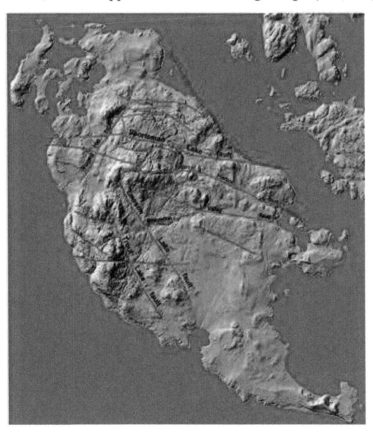

Figure 149. LIDAR image of San Juan Island showing major faults (red)

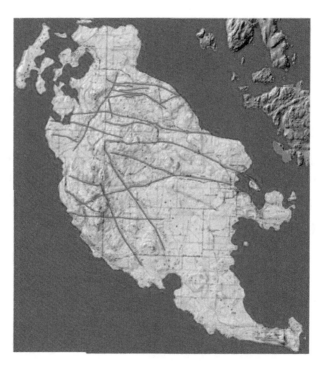

Figure 150. Topographic map of San Juan Island showing major faults (red)

SHAW ISLAND

LIDAR imagery shows that Shaw Island is cut by more than half a dozen high–angle faults (Fig. 151).

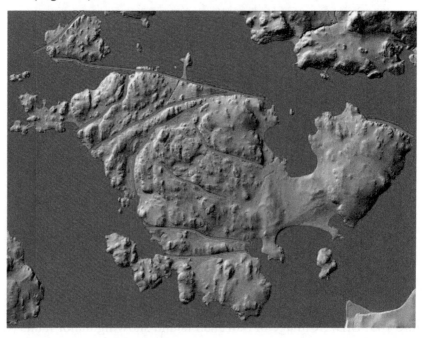

Figure 151. Faults (red) shown on LIDAR image of Shaw Island.

LOPEZ ISLAND

Southern Lopez Island is crossed by at least ten high–angle, faults (Figs. 152, 153). The northernmost of these faults has been mapped as the Lopez thrust fault and half a dozen parallel faults to the south were mapped as thrust faults. However, the linear straightness of the fault scarps, the near–vertical relief on the faults, and the linear trace of the faults across topographic relief indicates that they are really high–angle faults, not thrust faults. The straightness and near–vertical relief of several other parallel fault scarps to the south (Figs. 154–156) indicates that they are also high-angle faults. High angle shear zones also occur in the rocks on southern Lopez (Figure 157).

Figure 152. LIDAR image showing high–angle faults on southern Lopez Island.

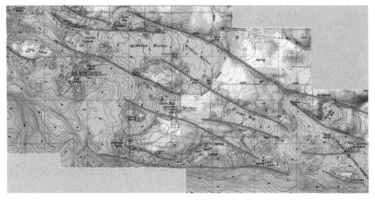

Figure 153. High–angle faults on southern Lopez Island.

Figure 154. The face of the cliff is a high–angle fault scarp. The arrow indicates the direction of upward movement of the fault along a nearly vertical fault plane. Southern Lopez Island

Figure 155. The face of the cliff is a high–angle fault scarp. The arrow indicates the direction of upward movement along the nearly vertical fault plane. Southern Lopez Island.

Figure 156. The face of the cliff is a high–angle fault scarp. The arrow indicates direction of movement along the nearly vertical fault plane. Southern Lopez.

Everson glaciomarine drift

Figure 157. High–angle fault zone, southern Lopez Island.

NEW DATA FROM SONAR IMAGERY

The geology of the sea floor in the San Juan Islands has remained virtually unknown until recently and has never been comprehensively mapped. The relationship of the sea floor geology to the surface geology of the islands has remained obscure despite sonar sea–floor imagery. Part of the problem seems to be that geologists didn't bother to do anything with the sonar data. In 2010, I began to take a serious look at the sea floor imagery and was amazed at what I found–many, many fault scarps, faults truncating geologic structures, folded beds, and perhaps most importantly, islands resting on broad submarine platforms bounded by large faults (Fig. 159). The islands we see above sea level are the tops of more extensive submerged geologic structures.

Interpretation of the sea floor geology from the sonar images depends largely on the sea floor geomorphology. Many fault scarps are apparent on the sonar images, but not all scarps are necessarily due to faulting. The main criteria for identifying fault scarps are (1) long, high, straight scarps with no other apparent origin, (2) truncation of geologic structures, and (3) long, linear gashes on the sea floor (4) scarps transverse to the direction of ice flow of the Cordilleran Ice Sheet.

Figure 158 shows a general summary of many (but not all) of the more prominent faults in the San Juan Islands. Many smaller faults are not shown but may be seen in later figures.

SAN JUAN–LOPEZ FAULT

A single, long, relatively straight fault scarp, mostly below sea level, makes the southwest margins of San Juan and Lopez Islands (Figs. 159, 160). Although other geologic processes can produce straight escarpments, they rarely do so for such long distances. A critical point here is that the San Juan–Lopez escarpment occurs well below sea level where normal subaerial surfaces processes are inoperative. The scarp is nearly at right angles to the direction of flow of the Cordilleran Ice sheet so it cannot be of glacial origin.

The San Juan–Lopez fault is the longest fault in the islands. It is at least 40 miles long, extending from the northern end of Stuart Island to beyond the SE corner of Lopez Island. (Figs. 159, 160) where it disappears beneath a submerged glacial moraine in Rosario Strait (Figs. 161–162). It appears to continue eastward through Deception Pass and into the North Cascade foothills, perhaps as the Devil's Mountain fault. The fault scarp reaches heights up to 1000 feet at the north end near Stuart Island and 300 feet at the southeastern corner of Lopez Island.

Figure 158. Sonar image of major high–angle fault scarps (red color) in the San Juan Islands. (Base map by NOAA)

Figure 159. San Juan–Lopez high–angle fault scarp. (Base map by NOAA)

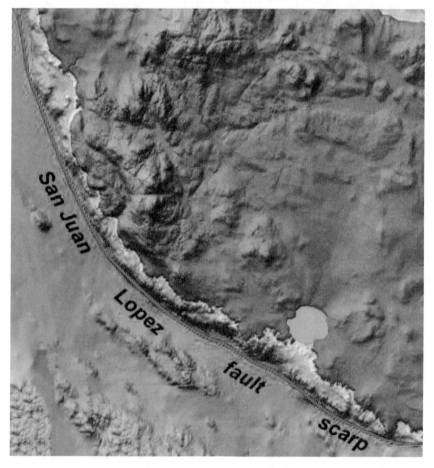

Figure 160. San Juan–Lopez high–angle fault scarp along the SW coast of San Juan Island. (Base map by NOAA)

The age of the fault is not accurately known. The topographic freshness of the scarp suggests that it is quite young. Pleistocene continental ice sheets over a mile thick crossed the islands more than half a dozen times, but the scarp is not severely eroded. A glacial moraine draped across the fault in Rosario Strait (Fig. 161) is not offset by the fault, so the fault must be older. Steep topographic relief on the scarp indicates that fault movement was down the dip of the fault plane.

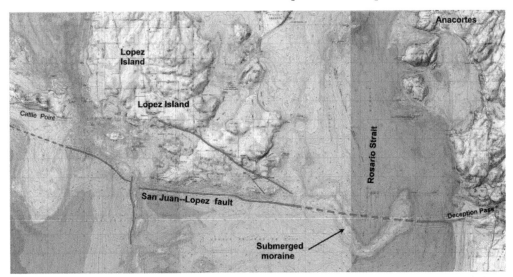

Figure 161. Submerged moraine draped across the San Juan–Lopez fault.

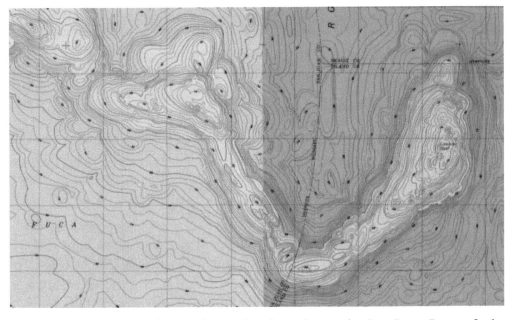

Figure 162. Submerged glacial moraine draped over the San Juan–Lopez fault about 500 feet above the sea floor at Lawson Reef in Rosario Strait. The glacier margin was banked against the north side of the moraine.

114

POINT CAUTION FAULT

The Pt. Caution fault extends the length of the NE coast of San Juan Island (Fig. 163). It may be even longer, but the bottom topography becomes complex both to the north and to the south. The height of the fault scarp is generally about 300 feet.

Figure 163. Point Caution and Shaw Island fault scarps, eastern San Juan Island. (Base map by NOAA)

SHAW ISLAND FAULT

The Shaw Island fault scarp makes the NE side of San Juan Channel (Fig. 163). The scarp is generally about 300 feet high and extends from Shaw Island to the Wasp Islands.

POINT LAWRENCE FAULT

The Point Lawrence fault scarp makes the entire northeast shoreline of Orcas Island from Point Lawrence to Point Thompson, a distance of 8 miles (Figs. 164, 165). The base of the scarp is ~350 feet below sea level and rises to ~400 feet above sea level NW of Point Lawrence. The southeast end of the scarp abruptly truncates the southern end of two sea floor ridges that are probably tilted beds of resistant Nanaimo or Chuckanut sandstone (Fig. 164). A submarine landslide at the base of the fault may be seen near the north end of the fault.

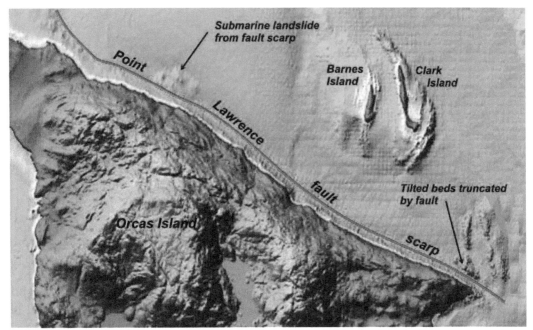

Figure 164. Point Lawrence high–angle fault scarp, NE shore of Orcas Island.
(Base map by NOAA)

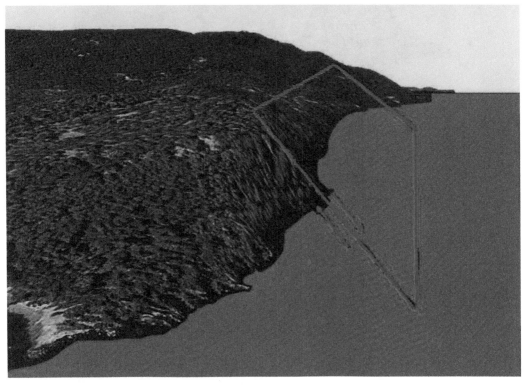

Figure 165. Point Lawrence high–angle fault scarp, northeastern Orcas Island.

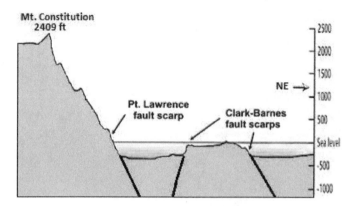

Figure 166. Cross section of the Point Lawrence fault, northeastern Orcas Island.

WEST BEACH FAULT

The West Beach fault makes the coastline along the NW shore of Orcas Island (Fig. 170). The subaerial scarp along the coast continues several hundred feet below sea level.

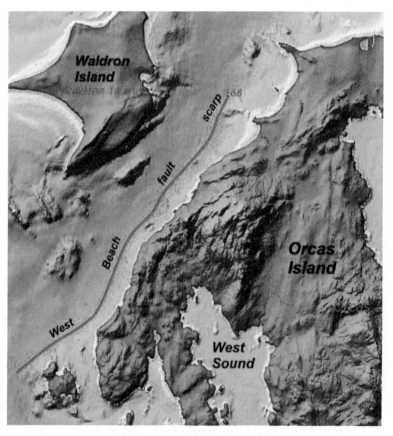

Figure 167. West Beach high-angle fault scarp, northwestern Orcas Island.
(Base map by NOAA))

SEA FLOOR FAULTS IN THE NORTHERN SAN JUAN ISLANDS

Many fault scarps are apparent on sonar imagery of the sea floor in the northern San Juan Islands (Figure 168). The faults show up as gashes on the sea floor, truncation of geologic structures, and linear scarps 300 to 1,000 feet high.

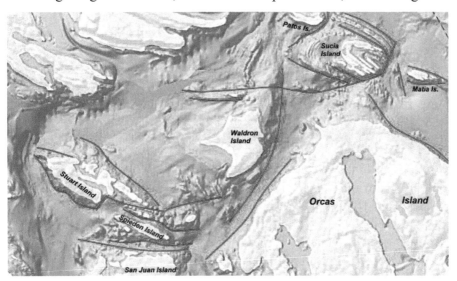

Figure 168. High–angle sea floor faults in the northern San Juan Islands. Red lines are faults. (Base map by NOAA)

SUCIA FAULT

The Sucia fault makes a long, conspicuous, east–west–trending gash on the sea floor westward from Sucia to beyond Skipjack Island. It truncates beds on the southern flank of the Sucia syncline and tilted beds north of Skipjack Island (Fig. 169). I first recognized this fault on sea floor imagery in 2000 and named it the Sucia fault. It lies on the same trend as the Vedder Mt. fault, which has a prominent linear, NE–trending scarp that can be traced for many miles along Vedder Mt. near Sumas in northeastern Whatcom County and into southwestern BC. Bedrock has been down–dropped several thousand feet and buried beneath thick Ice Age sediment near Sumas. Although the Vedder Mt. and Sucia faults aren't visibly connected, their alignment suggests that they may be the same fault.

OTHER FAULTS NEAR SUCIA ISLAND

Tilted beds of Nanaimo and Chuckanut sandstone at the SE end of the Sucia syncline are truncated by the Echo Bay fault (Fig. 170). The Echo Bay fault is a relatively short fault transverse to the Sucia syncline and Sucia fault. The southern terminus of the fault intersects the Sucia fault SE of Finger Islands in Echo Bay. The northern end of the fault disappears into the deep basin north of Sucia Island.

118

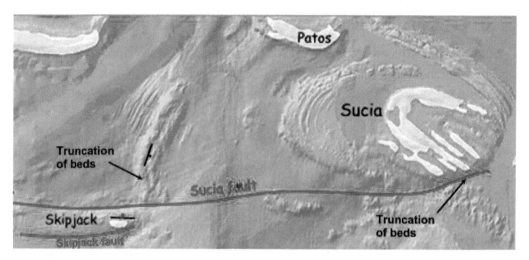

Figure 169. Sea floor sonar image of the Sucia fault. Note how the fault truncates the dipping beds of the Sucia syncline and the N–S trending linear ridges north of Skipjack Island. (Base map by NOAA)

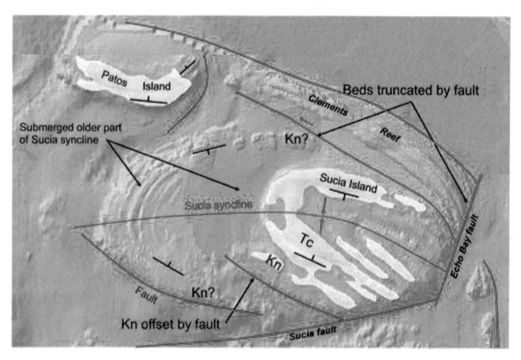

Figure 170. Sonar image of sea floor faults near Sucia Island. (Base map by NOAA)

Two other short faults offset beds of Nanaimo sandstone and conglomerate on the south limb of the Sucia syncline (Fig. 170). Most of the Nanaimo Formation makes up the western part of the submerged Sucia platform, which is not visible above sea level. A prominent ridge of Nanaimo sandstone making up the northern

119

limb of the syncline is truncated by a fault at a low angle to the trend of the bed (Fig. 170). This fault, which roughly parallels beds of Chuckanut sandstone making up Ewing Island and the sandstone ridge to the west, is truncated by the Echo Bay fault.

Clements Reef is composed of mostly submerged tilted beds (Fig. 171) that extend from the eastern end of the Sucia syncline almost to Patos Island. Beds making up the top of the reef are exposed at low tide and appear to dip steeply to the south, the same as the northern limb of the Sucia syncline just south of the reef (Fig. 171). Thus, the beds that make Clements Reef are apparently the same beds that make up the northernmost limb of the Sucia syncline, but have been moved westward by faulting (Fig. 170).

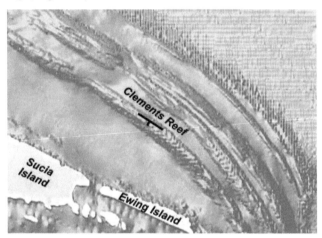

Figure 171. Clements Reef, made of SW–dipping Nanaimo sandstone beds. Everything with blue color is submerged below sea level.

PATOS FAULT

Patos Island is roughly aligned with Clements Reef, but the sandstone beds that make up the Island dip northward, the opposite direction as the beds making up Clements Reef (Fig. 170). Thus, they must be separated from Clements Reef by a fault.

RELATIONSHIP OF HIGH–ANGLE FAULTS TO MAPPED SAN JUAN THRUST FAULTS

The concept of the San Juan Thrust (Nappe) System is based on five postulated, single–plane, thrust sheets that have shoved large masses of rock very long distances, perhaps hundreds of miles. Figure 173 is redrawn from a recently published (2012) map proposing connection of the postulated San Juan thrust sheets to the North Cascades about a hundred miles to the east. The orange color represents the hypothesized Orcas thrust sheet and the green color represents the postulated Lopez thrust sheet. Arrows indicate the direction of assumed movement. However, as shown by LIDAR and sonar imagery, the distal portions of these postulated thrust sheets in the San Juan Islands are actually <u>not</u> thrust faults, but multiple segments of high–angle faults. The implications of the new LIDAR and sonar imagery are very significant because they invalidate the five thrust faults which form the basis for the San Juan Thrust System.

The validity of each of the proposed thrust sheets is discussed below. As shown by the new LIDAR and sonar imagery, all of the thrust faults are really segments of high–angle faults that have moved down the dips of the fault planes, rather than for extensive horizontal distances along low–angle thrust faults.

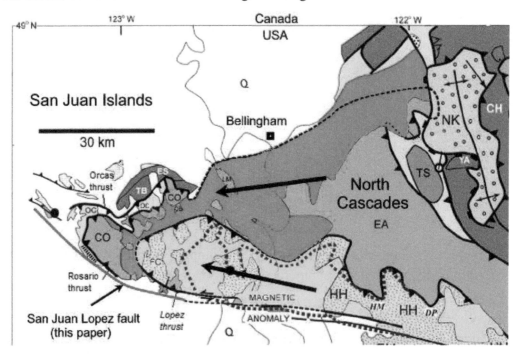

Figure 172. Postulated source of Orcas, Rosario, and Lopez thrust faults of the San Juan Thrust System, which are assumed to have moved great distances from the North Cascades (arrows). However, LIDAR and sonar imagery shows that these are actually high–angle faults, not thrust faults. The red line is the San Juan–Lopez high–angle fault, which lines up with the fault to the east. (Based on Brown, 2012)

VALIDITY OF THE ROSARIO THRUST FAULT

The Rosario thrust fault has been mapped as a single–plane, continuous, thrust fault that extends from the eastern tip of Orcas Island at Point Lawrence to Cattle Point on southern San Juan Island (Figs. 146). Figure 173 shows the relationship between the postulated Rosario/Orcas thrust sheet and high–angle faults between

Pt. Lawrence and Mt. Constitution. It's a rather complicated map, but what it shows is that the eastern portion of the Rosario thrust mapped on Orcas Island follows segments of several high–angle faults that are apparent on LIDAR images for many miles and it really isn't a thrust fault at all. LIDAR images show that the contacts between rock units have been incorrectly mapped, are offset along high–angle faults, and most of the Rosario/Orcas thrust sheet between Pt. Lawrence and Mt. Constitution consists of segments of high angle–faults, not a thrust fault.

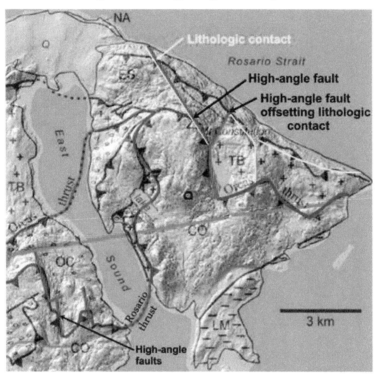

Figure 173. Portions of the Rosario and Orcas mapped thrust faults that are actually segments of high–angle faults.

A portion of the Rosario thrust is mapped crossing West Sound, bending around Crane Island at a sharp angle, and trending southeastward along the west shore of Shaw Island where it crosses several high–angle faults at right angles (Fig. 174). It intersects a high–angle fault at the southern end of Shaw Island, makes a sharp bend to briefly follow the high–angle fault, and crosses San Juan Channel. It then follows the Point Caution high–angle fault along the east coast of San Juan Island before diverging westward across San Juan Island where it follows the trace of several other high–angle faults. Thus, much of this portion of the so–called Rosario thrust really consists of segments of high–angle faults and is not actually a thrust fault.

The westernmost Rosario thrust as mapped follows the Point Caution high–angle fault along the northeast coast of San Juan Island (Fig. 174), then bends westward where it follows another high–angle fault across the northern end of the island. The mapped thrust then turns sharply to the south and wends its way along and among several high–angle faults to the southern part of the island (Fig. 175). For much of its extent, the postulated Rosario thrust actually consists of several segments of high–angle faults and is not a thrust fault.

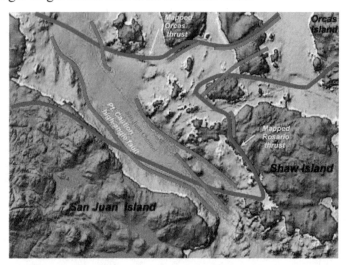

Figure 174. Relationship of the mapped Rosario thrust to the Pt. Caution fault. Blue=Rosario thrust; red=high–angle faults. (Base map by NOAA)

Figure 175. Relationship of Rosario thrust to high–angle faults on San Juan Island. Blue = Rosario thrust as mapped; red = high–angle faults.

VALIDITY OF THE ORCAS THRUST FAULT

The Orcas thrust has been shown on geologic maps from Point Lawrence across Orcas Island to West Sound and extending westward across San Juan Channel between Orcas and San Juan Island (Figs. 146, 176). However, much of the mapped thrust follows a linear trend across areas of topographic relief, indicating that most of it is actually a high–angle fault rather than a low-angle thrust fault.

The Orcas thrust has been mapped as merging with the Rosario thrust between Mt. Constitution and Pt. Lawrence, so the same issues as discussed for the Rosario thrust above also apply to the Orcas thrust, i.e., its trace from Mt. Constitution to Pt. Lawrence is not a thrust fault, but rather segments of high–angle faults.

Geologic maps show the trace of the Orcas thrust crossing the island between East Sound and the westernmost part of the island then across San Juan channel to the northern tip of San Juan Island.

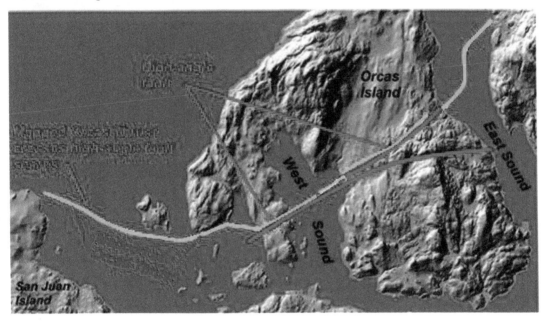

Figure 176. The Orcas thrust fault as mapped consists of several segments of high–angle faults. The mapped Orcas thrust fault is shown in green and the high–angle faults are shown in red. The postulated Orcas fault from East Sound to West Sound is actually a segment of a high–angle fault (Fig. 177). The western portion of the postulated Orcas thrust crosses several high–angle faults.

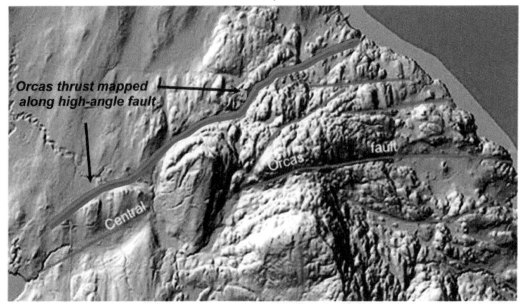

Figure 177. This segment of the Orcas thrust fault (blue) is actually a high–angle fault (top red) and is not a thrust fault. Several other high–angle faults parallel the high–angle fault that follows the postulated Orcas thrust.

Figure 176 shows the trace of the mapped Orcas thrust fault across western Orcas Island and San Juan Channel to San Juan Island. The Orcas thrust is mapped across the peninsula between East Sound and West Sound along the trace of a high–angle fault just north of the Central Orcas fault parallel to several other high–angle faults (Fig. 177). It then is shown crossing West Sound and following the prominent high-angle Central Orcas fault across the peninsula between West Sound and Deer Harbor (Fig. 176). The remainder of the Orcas thrust is mapped almost entirely on the sea floor as crossing the high-angle fault west of Jones Island, crossing San Juan Channel, and crossing the Point Caution high-angle fault on the east coast of San Juan Island. Thus, the Orcas thrust fault does not exist.

VALIDITY OF THE HARO THRUST FAULT

The Haro thrust is mapped as a low-angle thrust fault that extends across the entire breadth of the San Juan Islands (Figs. 146, 178). The western part of the postulated Haro fault is shown on geologic maps as three separate faults, merging eastward as they cross President Channel to the northwest shore of Orcas Island. The Haro thrust then circumscribes the entire northern coast of Orcas Island and swings sharply around Clark and Barnes Islands. Because almost all of the postulated Haro thrust lies along the path of these high–angle faults, it is not a thrust fault at all.

125

Figure 178. The trace of the mapped Haro thrust (blue) follows several segments of high–angle faults, including the West Beach fault, the north Orcas fault, and the Pt. Lawrence fault (red). (Base map by NOAA)

VALIDITY OF THE LOPEZ THRUST FAULT

The Lopez thrust fault is shown on geologic maps as extending across southern Lopez Island (Figs. 146, 179). Many smaller thrust faults haave been maped in the islands south of the Lopez thrust (Fig. 179). However, all of these are high–angle faults (Fig. 180) with high, straight, parallel, scarps and are not thrust faults.

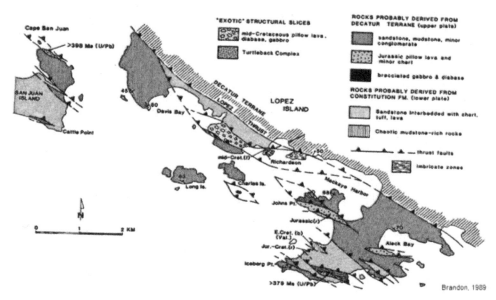

Figure 179. Geologic map of southern Lopez Island showing the Lopez thrust fault and all other faults mapped as thrust faults.

126

Figure 180. Mapped thrust faults (blue) and high–angle faults (red) on southern Lopez Island. The Lopez fault as mapped (blue) follows a high–angle fault (red) and is not a thrust fault at all.

BUCK BAY THRUST

The Buck Bay thrust fault is mapped as a single, low–angle, thrust fault that extends southward from northeast of Orcas Island down the entire length of Upright and San Juan Channels to Cattle Point (Figs. 146, 181,). However, the northeastern portion of the mapped Buck Bay thrust follows the prominent Olga high–angle fault (Fig. 181), the middle portion follows the scarp of a high-angle fault along the east coast of Shaw Island (Fig. 182), and the southern portion is submerged with no evidence of its existence. Thus, the Buck Bay thrust does not exist.

AGE OF THE HIGH–ANGLE FAULTS

Several lines of evidence suggest that two periods of high–angle faulting occurred in the San Juan Islands. Few of the faults above sea level shown on LIDAR have topographic scarps. Most show up on LIDAR as linear topographic gashes etched out by erosion, rather than scarps offsetting the land surface. Thus, they appear to be older than the prominent fault scarps at the margins of the larger islands and on the sea floor.

Many of the faults lie transverse to the flow of the Cordilleran Ice Sheet but do not show signs of glacial erosion. The San Juan–Lopez fault disappears beneath the end moraine on the sea floor in Rosario Strait without disturbing it, so it must be somewhat older than the last Ice Age (~20,000 to ~12,000 years ago).

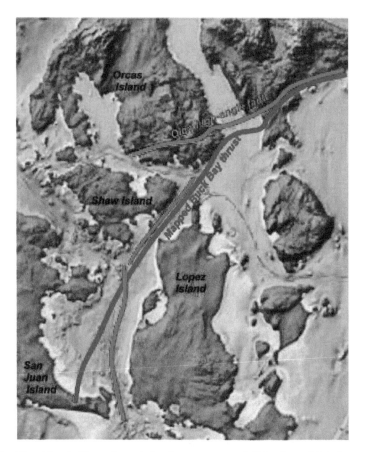

Figure 181. The Buck Bay thrust fault as mapped (blue) and high–angle faults (red). The mapped Buck Bay thrust fault follows several segments of high–angle faults and is not a thrust fault. (Base map by NOAA)

Figure 182. High–angle fault scarp mapped as the Buck Bay thrust fault on eastern Shaw Island. (Base map by NOAA)

The seafloor faults and island–bounding faults are marked by high, prominent linear fault scarps that cut across geologic structures and have not been affected by Ice Age glaciations even where they are transverse to the direction of ice flow of the mile-thick Cordilleran Ice Sheet, suggesting (but not proving) that the scarp is geologically young. The Point Lawrence fault scarp also has topographic evidence of landslides off the fault scarp (Figure 167).

However, a prominent, well-developed moraine at the south end of Rosario Strait is draped across the San Juan–Lopez fault, the longest and highest fault scarp in the islands. The morphology of the moraine is not offset by the fault so no movement has occurred on the fault since the moraine was deposited. The age of the moraine is not directly dated but must be slightly older than Everson glaciomarine sediments (11,700 ^{14}C yrs BP).

VALIDITY OF THE SAN JUAN THRUST (NAPPE) CONCEPT

LIDAR and sonar imagery clearly indicates that all five of the mapped thrust sheets that make up the San Juan Thrust (Nappe) System, the Rosario, Orcas, Haro, Lopez, and Buck Bay thrusts, are actually segments of high–angle faults and are not thrust faults. That is not to say that thrust faulting and shearing have not occurred–it simply means that shearing has been distributed throughout the pre-late Cretaceous rocks rather than along postulated large, single, discrete thrust sheets of large displacement (nappes). The geologic evidence for this conclusion is:

√   LIDAR imagery shows the traces of numerous high–angle faults on the islands.

√   Sonar imagery reveals many prominent, linear, high–angle fault scarps on the sea floor, many of which truncate geologic structures.

√   Most of the mapped Rosario thrust follows the trace of high-angle faults and is not a thrust fault.

√   Most of the mapped Orcas thrust follows the trace of high-angle faults and is not a thrust fault.

√   Most of the mapped Haro thrust follows the trace of high-angle faults and is not a thrust fault.

√   Most of the mapped Lopez thrust follows the trace of high-angle faults and is not a thrust fault.

√   Most of the Buck Bay thrust follows the trace of high-angle faults and is not a thrust fault.

Because none of the mapped thrust faults in the San Juan Islands are thrust faults, the San Juan Islands Thrust (Nappe) System does not exist and the stacked thrust (nappe) concept in the San Juan Islands is not valid.

GEOLOGY OF INDIVIDUAL ISLANDS

The San Juan Islands consist of three groups of islands (Fig. 183) having rather different geology: (1) the central islands composed of the oldest rocks in the islands, (2) the northern islands composed of the youngest rocks in the islands, and (3) the eastern islands composed of a mixture of igneous and metamorphic rocks.

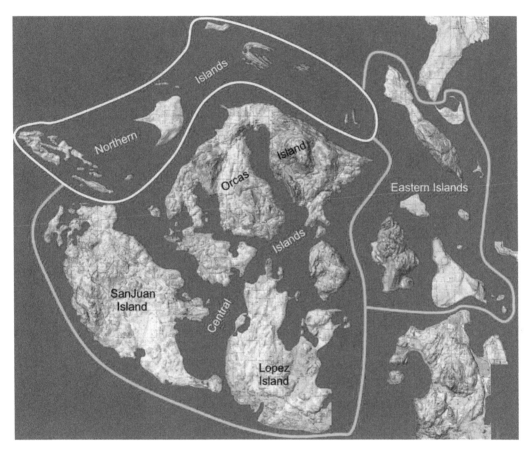

Figure 183. Northern, central, and eastern regions in the San Juan Islands.

CENTRAL ISLANDS

The central San Juan Islands include Orcas, San Juan, Lopez, Shaw, Blakely, Decatur, and numerous small islands (Fig. 183). These islands contain the oldest rocks in the islands and include a diverse assemblage of rocks ranging in age from Pre-Cambrian to Cretaceous.

ORCAS ISLAND

The topographic relief of Orcas Island is the most rugged of all of the islands (Fig. 184). Mt. Constitution rises to an elevation of about 2400 feet above sea level. Other high points include Buck Mt., Pickett Mt., Turtleback Mt., and Mt. Woolard. The geology of Orcas Island is quite diverse in both the types of rock present and in their age. A wide variety of sedimentary, igneous, and metamorphic rocks make up the island (Fig. 184).

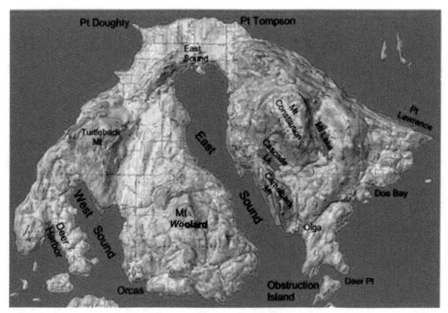

Figure 184. Orcas Island

The oldest rocks on Orcas Island are igneous and metamorphic rocks of the Turtleback Crystalline Basement Complex (see p.14-20). The two largest areas of Turtleback rocks are from Mountain Lake to Point Lawrence (Fig. 185) and Turtleback Mt. (Fig. 185, 187). Smaller wedges of Turtleback complex occur elsewhere on the island.

The Turtleback is overlain by Paleozoic marine greywacke sandstone, argillite, pillow basalt, and limestone lenses of the East Sound Group on the northern part of eastern Orcas Island near Buck Mt., along the entire NE coast between Pt. Lawrence and Pt. Thompson, and in a long narrow belt extending from East Sound to the southwest extremity of Orcas Island (Fig. 185). A basal conglomerate of the East Sound Group exposed along the shore near the foot of Orcas Knob is overlain by 25–50 feet of alternating beds of sandstone and argillite containing Devonian fossils. Along the northeast shore of Orcas Island, from Point Lawrence to the foot of Buck Mountain, greywacke, shale, argillite, conglomerate, and limestone (Fig. 185) trend about N 60° W and dip about 60°SW.

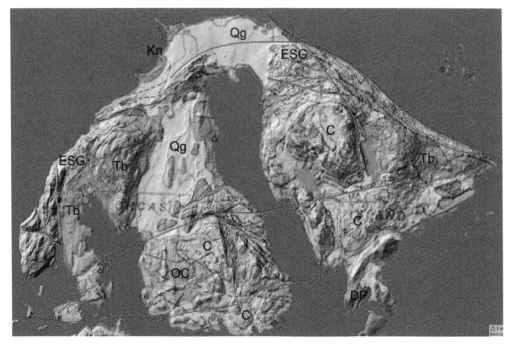

Figure 185. Revised geologic map of Orcas Island. Turtleback=Turtleback Crystalline Basement Complex. ESG=East Sound Group. OC=Orcas Chert. C=Constitution Fm. DP=Deer Point sandstone. Kn=Nanaimo sandstone. Qg=Ice Age glacial deposits. Red lines are high–angle faults.

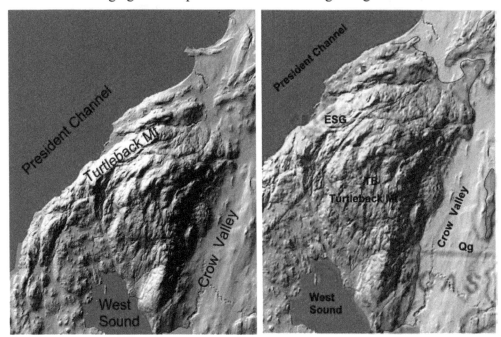

Figure 186. LIDAR image of Turtleback Mt. Figure 187. Geologic map of Turtleback Mt. TB=Turtleback Complex; ESG=East Sound Group; Qg=Ice Age glacial deposits.

In the Orcas lime quarry near the foot of Orcas Knob on Turtleback Mt., fossil brachiopods occur on weathered surfaces of limestone along the face of the cliff. The fossils have been etched into relief by greater chemical weathering of the enclosing limestone than the fossils, which are visible to a depth of only an inch or so, beneath which the limestone is entirely recrystallized and fossils are not visible. The brachiopod fossils are late Devonian *Atrypa reticularis*.

Orcas Chert is exposed along the west sides of Mount Entrance and Mt. Constitution and are overlain by greywacke. On Mt Constitution, the greywacke trends about N 35–45° E and dips SE at moderate angles.

Small limestones lenses containing micro-fossils of Carboniferous age occur intermittently from Raccoon Point to Point Lawrence (Fig. 188). Pebbles of interbedded greywacke and conglomerate beds are composed largely of Orcas chert, andesite, and volcanic tuff.

Figure 188. Limestone at Point Lawrence.

Paleozoic and Mesozoic rocks on Mt Constitution Range are nearly at right angles to sedimentary beds to the north indicating that the contact between them is a fault. Mountain Lake on the east flank of Mt. Constitution lies along the fault. The large Central Orcas fault crosses the entire island from Pt. Lawrence to Deer Harbor (Fig. 185). The rocks south of the fault near Olga consist of unaltered sandstone and shale of the Deer Point Formation, whereas the rocks to the north consist of altered greywacke and volcanic rocks of the Constitution Formation (Fig. 185). Beds of Deer Point sandstone and shale occur in gently undulating folds. Near its southwest entrance, Obstruction Pass occupies the axis of a syncline, while farther east it follows the axis of an anticline.

The southwestern peninsulas of Orcas Island (Fig. 189) are made up mostly of rocks of the Turtleback Crystalline Complex and marine sedimentary and volcanic rocks of the Easton Group, with smaller amounts of Orcas Chert (Figs. 190, 191). The contact between rocks of the Turtleback and East Sound Group is a straight line that cuts across variable relief, which means that the contact is essentially vertical. Several faults cut across the peninsulas on either side of Deer Harbor. The eastern fault is the continuation of the Central Orcas fault, a major fault that extends all the way to Point Lawrence.

The peninsula between West Sound and Deer Harbor is made of the Turtleback Crystalline Complex separated from Orcas Chert and the Constitution Fm. to the south by the Central Orcas Fault (Fig. 192)

Figure 189. The southwest peninsula of Orcas Island.

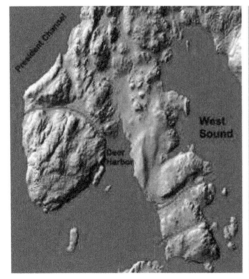

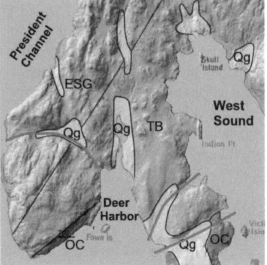

Figure 190. LIDAR image of the SW peninsula of Orcas Island. Red lines are faults.

Figure 191. Geologic map of the SW peninsula of Orcas Island. Red lines are Faults.TB=Turtleback, ESG=Est Sound Group, OC=Orcas Chert, Qg=Ice Age sediments

134

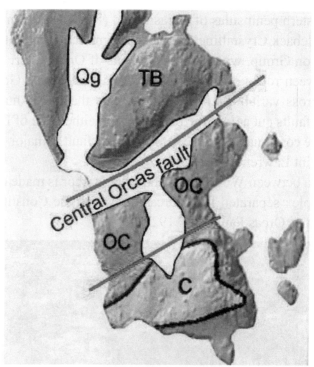

Figure 192. Geologic map of SW Orcas peninsula between West Sound and Deer Harbor. TB=Turtleback, OC=Orcas Chert, C=Constitution Fm., Qg=Ice Age sediments

The peninsula between East Sound and West Sound (Fig. 193) has rather complicated geologic structures. It consists of complexly faulted Orcas Chert and Constitution Formation, cut by the Central Orcas Fault and the Olga Fault, two very large faults that extend all the way across Orcas Island (Fig. 194). A high–angle fault separates these rocks from crystalline rocks of the Turtleback Complex to the north (Fig. 194). Mt. Woolard (Fig. 193), composed mostly of the Constitution Formation, is bounded on the east by a nearly vertical fault scarp that separates the Constitution Formation from Orcas Chert.

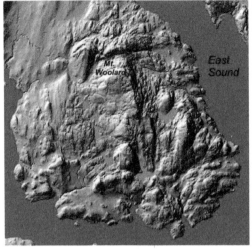

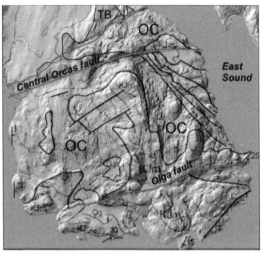

Figure 193. LIDAR image of the peninsula between East Sound and West Sound.

Figure 194. Geologic map of the peninsula between East Sound and West Sound. TB=Turtleback, OC=Orcas Chert, CF=Constitution Fm.

Nanaimo Formation, northshore of Orcas Island

The Nanaimo beds on the north shore of Orcas Island (Fig.195) consist of 3300 feet of conglomerate, sandstone, and shale which have been tilted and folded into anticlines and synclines (Figs. 196, 197). Differences in resistance to erosion of conglomerate and sandstone and shale control the topography along the shoreline. Conglomerate beds make rocky points and sandstone and shale have been etched out by erosion to form bays and beaches. Pebbles and cobbles in conglomerate beds consist mostly of chert, granite, and argillite.

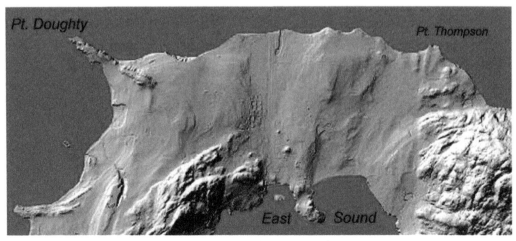

Figure 195. LIDAR image of the north shore of Orcas Island.

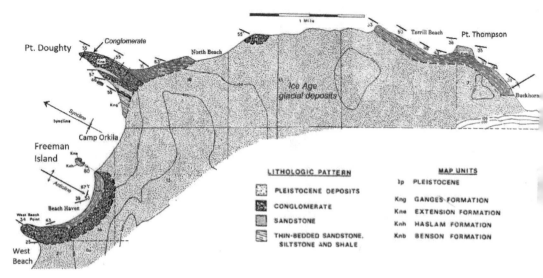

Figure 196. Geologic map of the north shore of Orcas Island.
(Modified from Carsten, 1982)

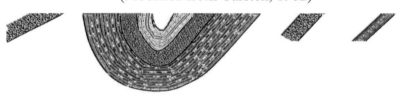

Figure 197. Geologic cross section of the northshore of Orcas Island from West
Beach to Point Doughty.

The Nanaimo beds have been subdivided into three units; the Haslam Formation
(mostly shale and sandstone), the Extension Formation (conglomerate), and the
Ganges Formation (mostly shale).

The Haslam Formation

The Haslam Formation consists of 1200 feet of shale and minor conglomerate
exposed along the shoreline from Beach Haven Resort to Freeman Point, from Point
Doughty to North Beach, and from Terrill Beach to Buckhorn. It is made up mostly
of thin-bedded sandstone and shale, massive sandstone, and conglomerate. The
upper part consists entirely of sandstone. Calcareous concretionary beds 2 to 18
inches thick that occur throughout the beds make thin resistance ridges on wave-
cut benches at the shoreline. Sandstone in the Haslam Formation is composed of
argillite (21%), chert (16%), granite and gneiss (7%), and volcanic rocks (6%).

Two sites contain abundant fossil shells in the Haslam Formation–the west side
of Point Thompson and west of North Beach Point (Fig. 73). Plant fossils are
common and are especially abundant in the upper sandstone.

137

The Extension Formation

The Extension Formation is 750 feet thick and consists mostly of conglomerate (Fig; 198) with smaller amounts of sandstone, and shale. Extension Formation conglomerate makes up Point Doughty, Freeman Island, Freeman Point and West Beach Point. The conglomerate lies on a scoured surface of the underlying Haslam Formation and is composed of pebbles and cobbles up to 10 inches in diameter, mostly granite and gneiss (23%) in the basal beds and 14% on the north shore of Point Doughty.

Lenses of coarse-grained sandstone occur about 65 feet above the basal contact and become increasingly thicker and more common higher in the section. The size of pebbles in the conglomerate decreases in the upper part and some shale is present.

Small petrified logs occur in some of the conglomerate beds, and plant fossils occur in shale in the upper part of the unit. The conglomerate and sandstone are strongly cemented and are resistant to erosion, forming points and ridges. The resistance to erosion increases with grain size, with the coarse beds being more resistant. Honeycomb weathering occurs in the coarse-grained sandstone and pebble conglomerate.

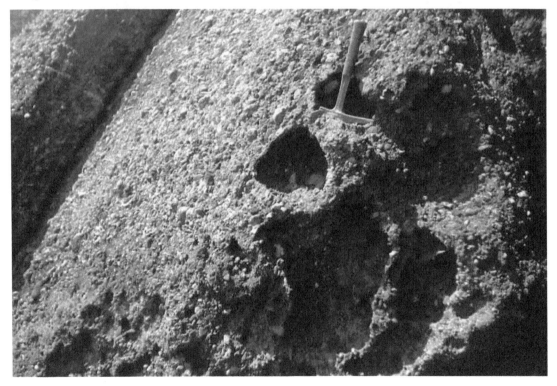

Figure 198. Steeply dipping conglomerate of the Nanaimo Extension Fm., north shore of Orcas Island.

The Ganges Formation

The Ganges Formation consists of at least 200 feet of predominantly shale with some sandstone, exposed only south of Point Doughty. The total thickness is not known because the top of the formation is nowhere exposed. The Ganges Formation is composed of grains of chert (29%), argillite (26%), quartz (13%), volcanic rock fragments (7%) and granite and gneiss (6%). A 13-foot-thick bed of coarse-grained sandstone in the upper part of the Ganges contains the oyster *Ostrea*.

Plant fossils are abundant in the Ganges Formation. Many pieces of logs (Fig. 199) that have turned to coal occur in organic shale. Weathering has etched out growth rings and other organic structures in many of the petrified logs. Calcite-cemented siltstone beds form small ridges on the wave-cut bench and often contain abundant leaf imprints.

Figure 199. Fossil tree in Nanaimo conglomerate, north shore of Orcas Island.

Source of the sediments

The dominant composition of pebbles and cobbles in the Nanaimo conglomerate include argillite, chert, granite, gneiss, and volcanic rocks that came from Orcas, Crane, Shaw, and San Juan Islands. The source of the argillite is most likely the Constitution Formation and Orcas Chert on Orcas and San Juan Islands. The source of chert was Orcas Chert. Granite, gabbro, and gneiss and are probably derived from the Turtleback Crystalline Complex on Orcas Island. The source of greywacke sandstone and fragmental volcanic rocks was probably Orcas and Jones Islands. Paleocurrent data from the Nanaimo Group beds exposed on Orcas Island indicate a likely source area to the south.

Point Doughty

Point Doughty (Fig. 200) is formed by beds of massive Extension Formation conglomerate (Fig. 201) that is more resistant to erosion than sandstone and shale beds above and below it (Fig. 201). To the south, conglomerate, sandstone, and shale, rich in fossil leaf impressions, overly the Point Doughty conglomerate.

A conglomerate bed that forms Freeman Island trends N55°W and dips 65–75° SW. Point Kimple, about ¾ of a mile south of Freeman Island, is composed of conglomerate with thin interbeds of sandstone and shale. trending N80°W and dipping 20–30°S. Nanaimo sandstone and shale trending N 65° W and dipping 60°N form a submarine shelf that extends from Orcas Island to Parker Reef offshore.

Figure 200. Point Doughty, north shore, Orcas Island. Conglomerate beds of the Nanaimo Extension Formation that form the point dip about 55°S (to the right)

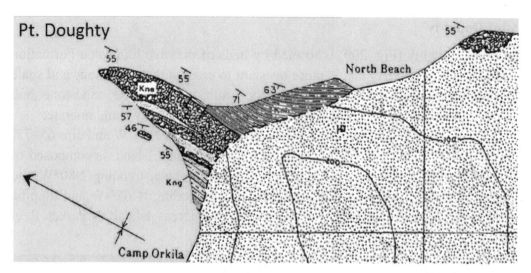

Figure 201. Geologic map of Point Doughty. Brown = Haslam Fm.; dark gray = Extension conglomerate; green = Ganges Fm. The beds dip about 55° S into a syncline whose axis goes through Camp Orkila. (Modified from Carsten, 1982)

West Beach Point

Conglomerate at West Beach Point (Fig. 202) dips to the south (right) at 25–40° (Fig. 196), forming the southern flank of an anticline whose axis lies between Freeman Island and West Beach Point.

Figure 202. West Beach Point, north shore Orcas Island. Note the rocky ridges of resistant sandstone that extend seaward from the end of the point.

Point Thompson

Massive, resistant sandstone in the lower part of the Haslam Formation that forms Point Thompson (Fig. 203) dips 40–60° S (Figs. 204, 205). Other massive sandstone beds occur east of Point Thompson, on Terrill Beach, and at North Beach Point (Fig. 204). Two conglomerate beds that are exposed just west of Buckhorn contain abundant vein quartz and shell fragments of fossil oysters (*Inoceramus, Ostrea*). A shark tooth was also found there.

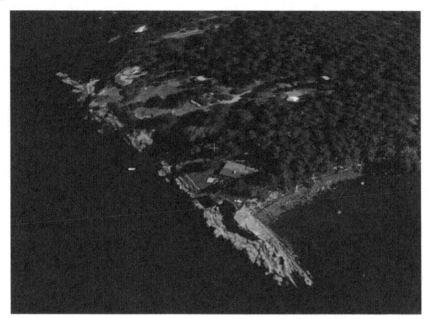

Figure 203. Point Thompson, Orcas Island. Nanaimo beds dip to the south (right).

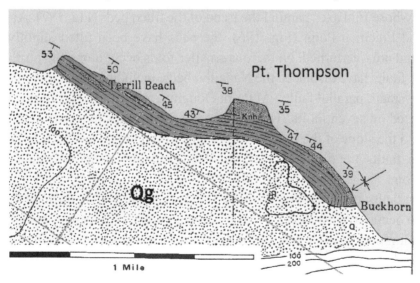

Figure 204. Geologic map of Point Thompson.

142

Figure 205. South dipping beds of Nanaimo sandstone making Point Thompson.

Folding and faulting of the Nanaimo beds

From Point Doughty to Buckhorn (Fig. 196), the Nanaimo beds are tilted to the south and trend northwestward approximately parallel to the shoreline. Between Point Doughty and West Beach, the beds have been folded into a syncline and an anticline whose fold axes parallel the trend of the tilted beds (Fig. 197). At Freeman Island and Freeman Point (Fig. 196), the beds have been tilted slightly beyond vertical and are overturned. Numerous smaller folds occur along the north shore.

Many faults have displaced the Nanaimo beds, including both large major faults and many small, parallel faults. At Point Doughty, at least 137 feet of displacement has occurred on seven faults. The trend of the fault planes ranges from N2O°W to N25°E and the slope of the fault planes ranges from vertical to 75 ° E to 80 ° W. At least five faults trending N5°–9°E displace Nanaimo conglomerate 67 feet at Airport Point.

Major faults trending N63°E to N75°E and dipping 67°S through vertical to 75°N also occur at West Beach Point, Freeman Point, and Freeman Island. Several of the fault planes exhibit zones of pulverized rock and surfaces polished by fault movement.

Places to look for fossils. Be sure to get permission from property owners.

A shoreline exposure of coarse–grained sandstone and conglomerate northwest of Buckhorn contain fragments of *Inoceramus* and *Ostrea.*

An exposure of shale on both the wave-cut bench and the cliff face on the west shore of Point Thompson about 250 feet south of the tip of the point contains fossil ammonites, *Inoceramus*, crustaceans, and echinoderms.

A seacliff exposure of shale on the west flank of Point Thompson contains *Inoceramus* in growth positions.

A seacliff exposure of shale about 800 feet west of North Beach Point contains *Inoceramus,* crustacea and ammonite fragments.

A seacliff exposure of shale south of Point Doughty about 400 feet north of a private dock contains abundant fossil plants, including ferns, conifers, palm fronds, and ginkos.

A seacliff exposure of sandstone south of Point Doughty about 130 feet north of a private dock contains rare *Ostrea* shells.

An ammonite fragment was found at a seacliff exposure of shale 800 feet south of Freeman point.

ICE AGE DEPOSITS

Everson glaciomarine stony silt

Almost all of the area landward from the north coast of Orcas is mantled with Everson glaciomarine stony silt and sand (Fig. 196). On the east side of Ship Bay west of Crescent Beach at the north end of East Sound, glaciomarine sediments containing many small pieces of wood rises from beach level to the top of the sea cliff about 50 feet above sea level (Fig. 206). It overlies bedded sand, which rests on silt and clay near beach level.

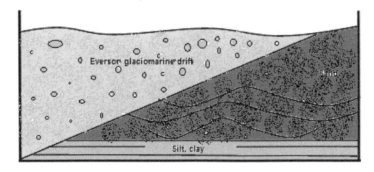

Figure 206. Geologic cross–section of sea cliffs at Ship Bay at the north end of East Sound. The glaciomarine drift is 11,700 ^{14}C yrs old.

Many fossil shells occur in about 20–30 feet of glaciomarine sediments at West Beach about 100 feet north of rock outcrops there. Glaciomarine stony silt and sand about halfway between Camp Orkila and Pt. Doughty contain shell fragments. At a road cut near Olga, 10–15 feet of glaciomarine stony silt containing a few scattered shell fragments is overlain by south–dipping beach sand

One of the richest fossil localities on Orcas, containing eight species of clams, occurs in glaciomarine stony silt and clay in a ditch beside a field next to the road two miles SW of East Sound. Shells there were radiocarbon dated 12,350 ± 400 ^{14}C years old. Abundant fossil shells occur in glaciomarine pebbly silt and clay at Four Winds Camp on the peninsula west of West Sound. The shells there have been radiocarbon dated at 12,600 ± 400 ^{14}C years old. These sediments are overlain by about 10 feet of very fossiliferous beach sand at an elevation of 125 feet.

LATE GLACIAL SHORELINES

Multiple marine terraces extend landward from the shoreline to elevations up to 300 feet above present sea level at many places on Orcas Island. They consist of numerous benches and former beach ridges and spits that are not easily seen from the ground or from the air, but show up in great detail on LIDAR imagery (Fig. 117, 207). Near the end of the last Ice Age, all of Orcas Island less than 300 feet above present sea level was submerged, and the island looked much different than it does now. In Figure 207, only the area above 300 feet (blue line) was above sea level. Figure 117 is a LIDAR image showing the area near Point Doughy in greater detail. Figure 208 shows late Ice Age shorelines south of Point Thompson. Figure 209 shows shorelines near Olga and Figure 210 shows shorelines along East Sound. Figure 211 shows what Orcas Island would have looked like about 14,000 years ago.

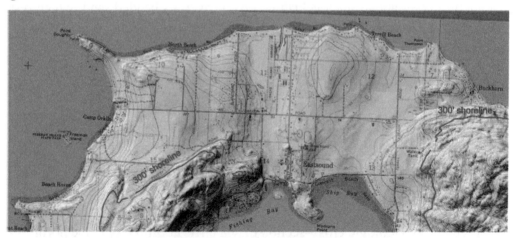

Figure 207. Lidar and topographic map of northern Orcas
showing raised marine shorelines at 300 feet.

145

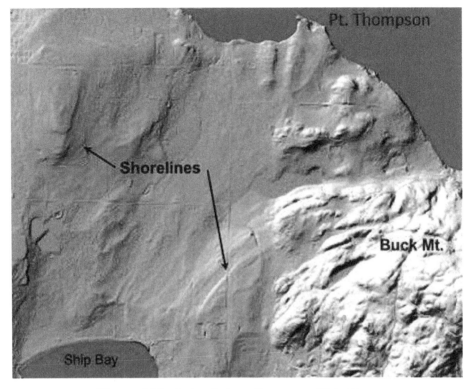

Figure 208. Raised shorelines near Ship Bay.

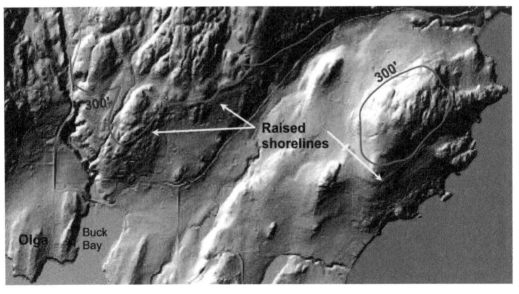

Figure 209. Raised shorelines at Olga on southern Orcas Island. The former shorelines extend to 300 feet above present sea level.

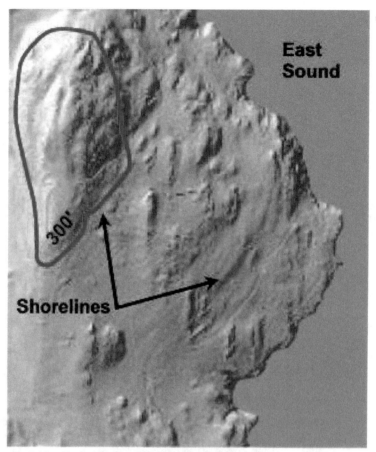

Figure 210. Raised shorelines along the west side of East Sound.

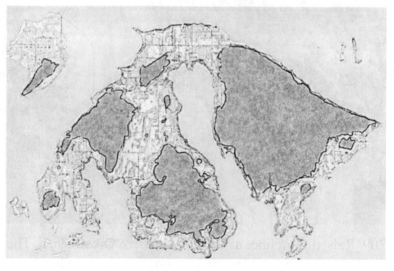

Figure 211. Orcas Island as it looked about 14,000 years ago when the shoreline was 300 feet higher than it is now. Orcas Island then consisted of three main islands (shown in brown) and a half dozen small ones. The blue area was submerged.

147

SMALL ISLANDS NEAR ORCAS ISLAND

OBSTRUCTION ISLAND

Greywacke and shale of the Deer Point Formation on Obstruction Island (Fig. 212) trends generally N45–50° E and dips 25° SE. At the eastern corner of the island, an anticline plunges eastward at a gentle angle

The highest point on the island is 240 feet above sea level so it was completely submerged during the late Ice Age 300–foot sea level stand. Just to the south, well–defined raised marine shorelines occur 300 feet above sea level on the north end of Blakely Island, so Obstruction Island would have lain more than 60 feet below sea level at that time.

Figure 212. Obstruction Island.

DOE ISLAND

Deer Point argillite and greywacke sandstone on Doe Island trend N70°E and dip SE at a gentle angle.

PEAPOD ROCKS

Peapod Rocks are composed of Deer Point greywacke sandstone, conglomerate, and argillite, which trend NE and dip to the SE. They belong to the same stratigraphic units as Deer Point and Obstruction Island.

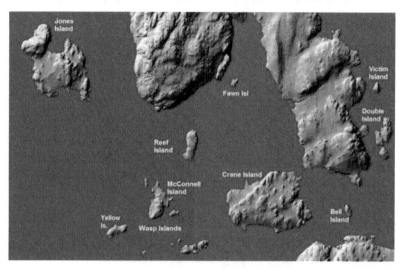

Figure 213. Small islands southwest of Orcas Island.

JONES ISLAND

Greywacke sandstone and limestone of the East Sound Group on Jones Island (Figs. 213, 214) have been intruded by dark igneous rocks. A 20–foot-thick limestone bed that trends northwesterly along the east shore of the island bends abruptly westward at the large bay on northern Jones Island and continues to the western coast of the island. The limestone, exposed on two small peninsulas and a small island along the shore, is interbedded with black cherty argillite and overlies volcanic rocks. Another limestone lens interbedded with volcanic rocks 1000 feet to the east is 14 feet thick and extends for 100 feet. It contains a few poorly preserved coral and bryozoan fossils that appear to be Devonian. A third limestone occurs along the east coast of the island about 800 feet farther south. It makes a ridge that protrudes from the beach gravel as a narrow, vertical bed with poorly preserved Devonian coral and crinoid fossils. Microscopic conodont fossils also indicate an early Devonian age (~400 million years).

Figure 214. Jones Island
149

CRANE ISLAND

Crane Island (Figs. 213, 215, 216) is the largest island of the Wasp Island group. It is composed of intensely deformed argillite and ribbon chert of the Constitution Formation underlain by volcanic rocks and limestone. The beds trend northeasterly and dip southward along the western limb of a syncline.

At a small peninsula near the southwest corner of the island, a limestone lens about 12 feet thick is apparently a continuation of the limestone on Cliff Island. A fault cuts across the northwestern part of the island and continues across the peninsula of Orcas Island parallel to the central Orcas fault (Fig. 216).

Figure 215. Crane Island.

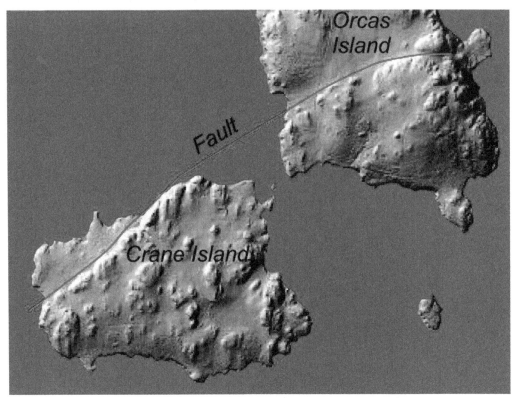

Figure 216. Crane Island. The red line is a fault that extends from Crane Island to the peninsula on Orcas Island west of West Bay.

150

VICTIM ISLAND

Orcas Chert on Victim Island (Fig. 217) trends N60°W and dips to the southwest, similar to rocks along the shore of Orcas Island to the west. The chert is cut by several small igneous intrusions.

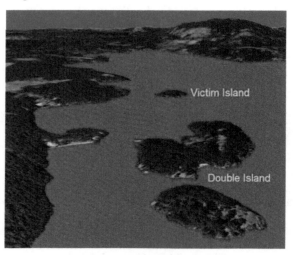

Figure 217. Double and Victim Islands, West Sound, Orcas Island

DOUBLE ISLAND

Double Island (Fig. 217) is composed of Orcas Chert having the same trend and dip as on Victim Island. The peninsula between West Sound and Deer Harbor is a minor syncline plunging to the south. The chert on Double Island and Victim Island are a part of the eastern limb of this fold.

BELL ISLAND

Bell Island (Fig. 218) consists of chert of the Constitution Formation on the axis of a syncline that makes up the peninsula to the north.

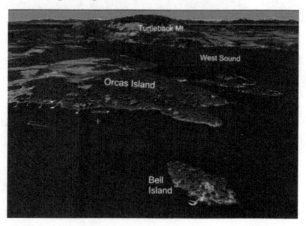

Figure 218. Bell Island.

CLIFF ISLAND

Cliff Island (Fig. 219) lies off the NW coast of Shaw Island SW of Crane Island. Orcas Chert on the island trends northeasterly parallel to the length of the island and dips steeply to the southeast. Several limestone lenses occur along the northwest shore of the island. Igneous intrusions cut the chert.

Figure 219A. Cliff Island NW of Shaw Island.

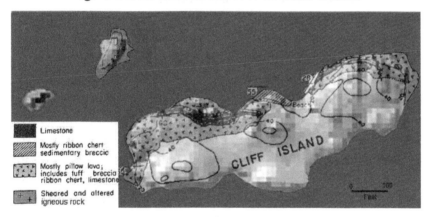

Figure 219B. Geologic map of Cliff Island. (Modified from Danner, 1966)

McCONNELL, YELLOW, COON, REEF, AND FAWN ISLANDS

McConnell, Yellow, Coon, Reef, and Fawn Islands (Fig. 221) all consist of rocks of the Turtleback Crystalline Complex.

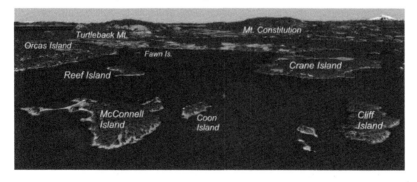

Figure 220. Wasp Islands between Orcas and Shaw Islands.

SAN JUAN ISLAND

Most of San Juan Island (Fig. 221) consists of greywacke sandstone, argillite, and volcanic rocks of the Constitution Formation (Fig. 222) with smaller areas of Orcas Chert, Turtleback Crystalline Complex, Garrison schist, and Haro Formation. Much of San Juan Island is mantled with Ice Age glaciomarine sediments up to 300 feet above present sea level and glacial till and outwash above 300 feet. All of the rock units except the Haro Formation at Davidson Head have been intensely sheared and show closely spaced shear planes (parallel planes along which movement has occurred, much like shuffling a deck of cards). They have also been affected by low–grade metamorphism and contain numerous quartz veins.

The contact between Orcas chert and overlying greywacke sandstone of the Constitution Formation is exposed along the northeast shore of San Juan Island about a mile east of Sportsmans Lake. Orcas Chert occurs at the northwest end of the island northwest of the San Juan Range, at Mt. Dallas and at the southeast side of False Bay. Along the southwest margin of San Juan Island, the chert trends parallel to the shoreline and dips to the northeast. Along the northeast margin of the island, the chert dips to the southwest. On the northwest end of the island geologic structures are more complex as a result of folding and by faulting.

Greywacke at the base of the stratigraphic section is exposed near Point Caution along the northeast shore of the island, on Biological Hill, the San Juan Range, Mt. Grant, and Little Mountain. Because the greywacke sandstone is more resistant to erosion than overlying and underlying beds, it forms ridges. The greywacke is overlain by about 100 feet of coarse conglomerate at Friday Harbor and on the southwestern part of the island. The conglomerate is overlain by several hundred feet of massive argillite with interbeds of greywacke and conglomerate.

Figure 221. LIDAR image of San Juan Island.

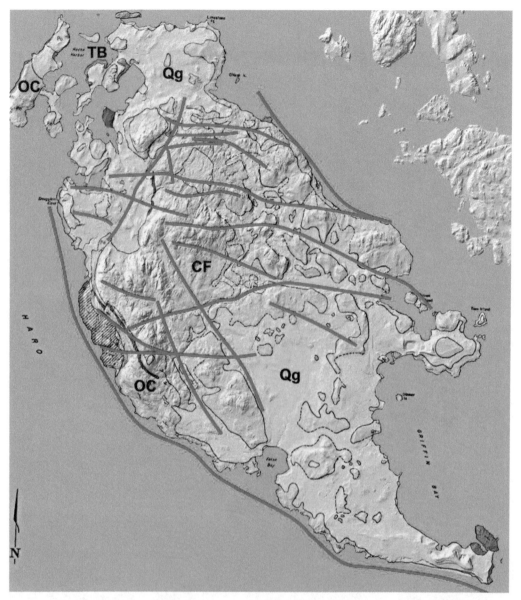

Figure 222. Geologic map of San Juan Island. TB = Turtleback Complex, OC = Orcas Chert, CF = Constitution Fm., Qg = glacial sediments; red lines are faults.

Lime Kiln Lighthouse Point

Pillow basalt and volcanic breccia with interbedded limestone lenses and ribbon chert occur along the west coast of San Juan Island from Deadman Bay to Smallpox Bay (Fig. 223). The limestone (Fig. 224) has been recrystallized to marble in irregular lenses interbedded in the volcanic rocks. Numerous microscopic fusulinid and conodont fossils in the limestone lenses indicate a Permian age. Microscopic radiolarian and conodont fossils in ribbon chert range in age from Permian to Triassic.

155

The largest limestone lens was the site of the Crowell quarry and kiln (Fig. 225), after which Lime Kiln Point was named. The limestone there consists of irregular, steeply dipping lenses interbedded with pillow basalt and smaller amounts of ribbon chert, greywacke sandstone, and argillite.

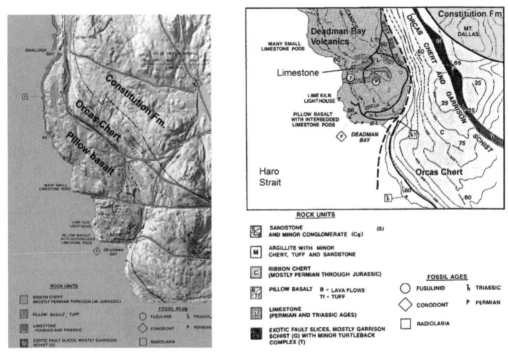

Figure 223. Geologic maps of Lime Kiln Point

Figure 224. Permian limestone lens in volcanic rocks
at Deadman Bay, San Juan Island.

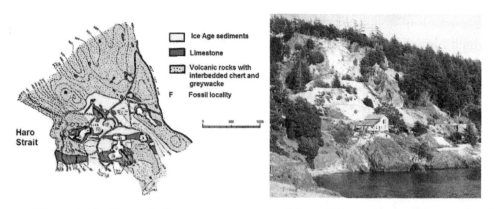

Figure 225. Crowell limestone quarry at Lime Kiln Point, San Juan Island.

Lime Kiln Point is a famous whale watching site that is geologically important because of good exposures of limestone containing Permian microfossils (fusulinids). The limestone has been recrystallized to marble whose age ranges from Early Permian to Late Triassic, based on fusulinid, conodont, and radiolarian microfossils.

At least five small lenses of limestone are interbedded with volcanic rocks at Deadman Bay. Most of the limestone beds are less than 10 feet thick. The north side of Deadman Bay consists mostly of basalt and volcanic breccia interbedded with ribbon chert, argillite, and limestone dipping steeply to the northeast. Southeast of Deadman Bay, most of the rocks exposed along the shore consist of ribbon chert. The largest limestone exposure occurs close to the West Side road near its junction with the road to the Lime Kiln lighthouse. Another limestone bed about 25 feet long and 3 feet thick occurs along the road near the southwest end of the area.

Roche Harbor

Rocks in the Roche Harbor area consists of the Turtleback Crystalline Complex, Devonian conglomerate, Roche Harbor Permian limestone, Orcas Chert, Haro Formation, and Ice Age sediments (Fig. 226). Turtleback amphibolite, gneiss, and porphyry (TB) makes up Bazalgette Point and gabbro and pyroxenite occur in small lenses here and there. Turtleback phyllite makes up Bell Point across Westcott Bay from the limestone quarries. Orcas chert (OC) makes up all of Henry Island.

The Roche Harbor limestone consists of several layers of Permian limestone interbedded with ribbon chert, greywacke sandstone, argillite, and volcanic rocks (Fig. 227) that have been folded into several NW–trending, intensely deformed anticlines and synclines. It was extensively quarried for many years and the abandoned pits (Fig. 228) now extend along most of the peninsula from the town of Roche Harbor to Mitchell Bay.

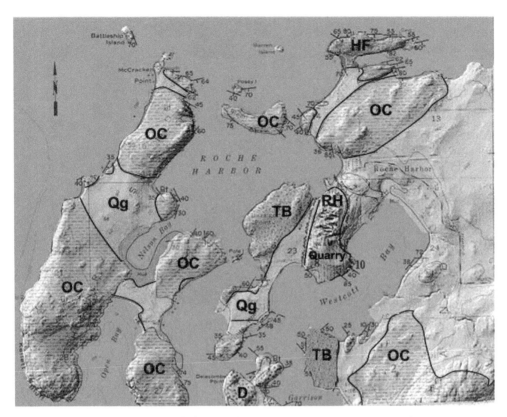

Figure 226. Geologic map of the Roche Harbor area. TB=Turtleback Crystalline Complex, D=Devonian conglomerate, OC=Orcas Chert, RH=Roche Harbor Permian limestone, HF=Haro Formation, Qg=Ice age and recent sediments. (Modified from Danner, 1966)

Figure 227. Limestone lens in chert at Roche Harbor. (Photo by Ted Danner)

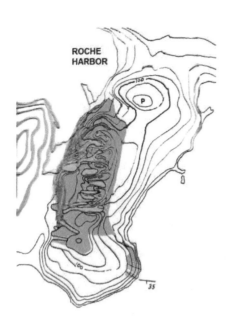

Figure 228. Limestone quarries (blue) at Roche Harbor.

Limestone Point

Limestone Point is a small knob of limestone extending into Spieden Channel on the northeast tip of San Juan Island (Fig. 229. The limestone consists of tiny microscopic spheres known as oolites (Fig. 230), and larger fragments of mollusks, gastropods (marine snails), brachiopods, and echinoderms that occur as thin layers interbedded with chert. South of the quarry, the limestone is interbedded with chert and pinches out laterally. Beds in the quarry trend NE and dip steeply to the northwest.

Figure 229. Limestone Point, NE tip of San Juan Island.

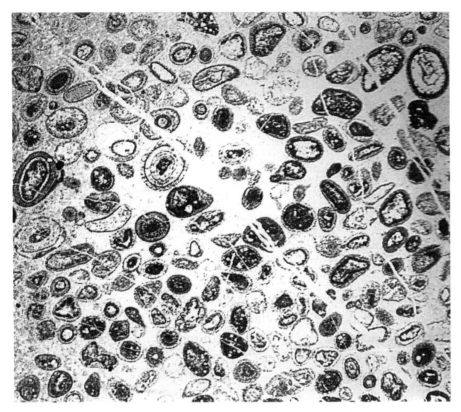

Figure 230. Microscopic view of oolitic limestone at Limestone Point,
NW San Juan Island.

Eagle Point

At Eagle Point, small pods of Garrison schist occur near the base of a sequence of Orcas ribbon chert, pillow basalt, and argillite and as small fault slices in Orcas Chert (Fig. 231). Small limestone pods in argillite at Eagle Cove contain late Triassic conodont fossils and Orcas Chert at Eagle Point contains Triassic radiolarian fossils. Massive greywacke sandstone of the Constitution Formation overlies Orcas Chert (Fig. 231). Pebbles of Garrison schist and Orcas Chert in the basal sediments of the Constitution Formation indicate that the Constitution Formation was deposited on the Orcas Chert, similar to the situation on Orcas Island.

Garrison schist varies in thickness from a few meters to a few tens of meters. The rocks are generally green, fine-grained schists with minor quartz-mica schist and minor recrystalized limestone. Well-developed schistosity (parallel alignment of mineral grains) of the Garrison rocks is different than less-metamorphosed rocks of Orcas Chert and Constitution Formation. Isotope dates of 286 ± 20, 242 ± 14, and 167 ± 12 million years have been obtained from the schist.

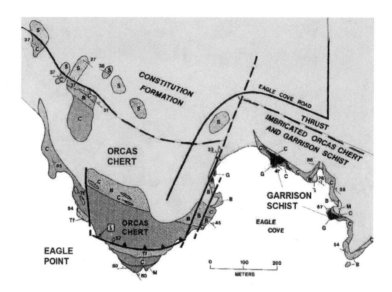

Figure 231. Geologic map of Eagle Point, SW San Juan Island. Pods of Garrison schist occur associated with Orcas chert, pillow basalt, and argillite, which is overlain by greywacke sandstone of the Constitution Formation. (Modified from Brandon et al., 1988)

South Beach

Near South Beach on SW San Juan Island, several small pods of Garrison schist occur near the shorelines and are overlain by Orcas chert with small fault slices of schist (Fig. 232). Orcas Chert is overlain by greywacke sandstone of the Constitution Formation.

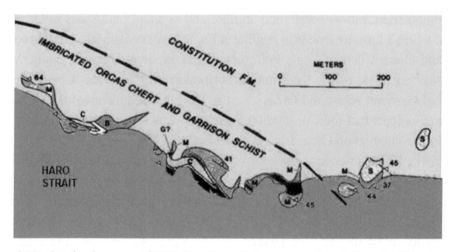

Figure 232. Geologic map of SW San Juan Island near South Beach. Black= Garrison schist, brown=argillite, gray=pillow basalt. (Modified from Brandon et al., 1988)

Cattle Point

Cattle Point lies at the southeastern tip of San Juan Island (Fig. 233). Near sea level, the point consists of highly sheared greywacke sandstone, argillite, and chert with many quartz veins (Fig. 234). These rocks are overlain by glacial outwash sand and capped with glaciomarine pebbly silt containing marine fossils.

Figure 233. Cattle Point, San Juan Island. The rocky point near sea level consists of greywacke sandstone, argillite, and chert overlain by Ice Age sediments. The sea cliff on the left side is made of glacial outwash sand (light color), overlain by slightly darker glaciomarine pebbly silt.

Figure 234. Greywacke sandstone cut by numerous quartz veins, Cattle Point.

Ice Age Features on San Juan Island.

During the last Ice Age about 14–17,000 years ago, the San Juan Islands were overrun by an ice sheet from British Columbia about 6,000 feet thick (Fig. 96). The best exposures of Ice Age sediments and glacial erosion features are at Cattle Point (Fig. 233). The rocks near sea level at Cattle Point have been glacially polished and grooved (Fig. 97, 98) and are overlain by non–marine stream sediments (Fig. 235) and glaciomarine pebbly silt (Fig. 236).

15,000 years ago, abrupt global warming of more than 25 °F in a century caused massive melting of ice sheets all over the world. As the ice thinned and retreated, the ice margin lingered for a while near Friday Harbor where meltwater flowed into the sea and built a marine delta just south of Friday Harbor, radiocarbon dated at $13,240 \pm 70$ ^{14}C years ago (about 16,000 calendar years) (Figs. 237). The gravel pit (Figs. 115, 238) is now largely mined out. A long sinuous former spit extends eastward from Friday Harbor (Fig. 115) and half a dozen lower shorelines occur between the raised spit and present sea level.

The ice sheet over the San Juan Island melted rapidly, thinned drastically, and collapsed, leaving the entire lowland covered with ice bergs floating in a sea that was 300 feet higher than present. A layer of massive pebbly silt 10-20 feet thick, known as glaciomarine drift, was deposited from the floating ice over all of the islands up to 300 feet above present sea level. These deposits commonly have numerous marine shell fossils that have been radiocarbon dated at 11,000 to 12,500 years old. Good examples of these fossil-bearing sediments can be found in many places. On San Juan Island, among the most richly fossiliferous localities are at Cattle Point and Davidson Head where glaciomarine sediments that contain abundant marine fossils radiocarbon dated at $12,160 \pm 290$ ^{14}C years ago (about 14,000 calendar years) (Fig. 239).

Figure 235. Ice Age non–marine sand and gravel overlain by glaciomarine pebbly silt at Cattle Point.

164

Figure 236. Glaciomarine pebbly silt at Cattle Point.

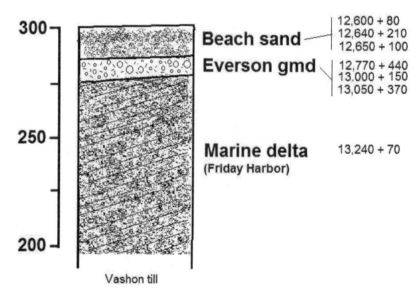

Figure 237. Generalized stratigraphic section of youngest Ice Age deposits.

Figure 238. Gravel pit in Ice Age marine delta at Friday Harbor.

Figure 239. Fossil shells in glaciomarine sediments at Davidson Head.

Late Ice Age marine terraces

When the floating ice had melted away, marine terraces and beaches were built 300 feet above level sea level and as sea level successively lowered, numerous marine terraces and beach ridges were left on almost all of the islands. Marine terraces at Cattle Point extend to the top of Mt. Finlayson 300 feet above present sea level (Fig 113, 114). Numerous lower marine terraces and long sinuous former spits occur westward to False Bay (Fig. 240). American Camp is located on these terraces. San Juan Island had a much different appearance at that time (Fig. 241).

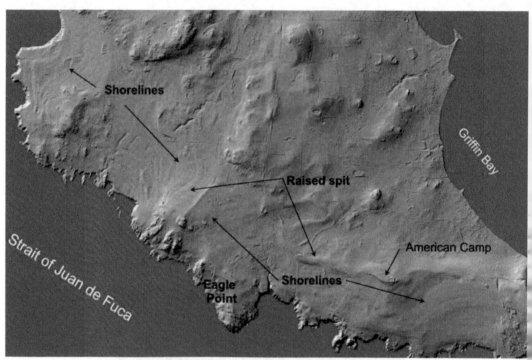

Figure 240. Lidar image of marine terraces and raised spits between Cattle Point and False Bay.

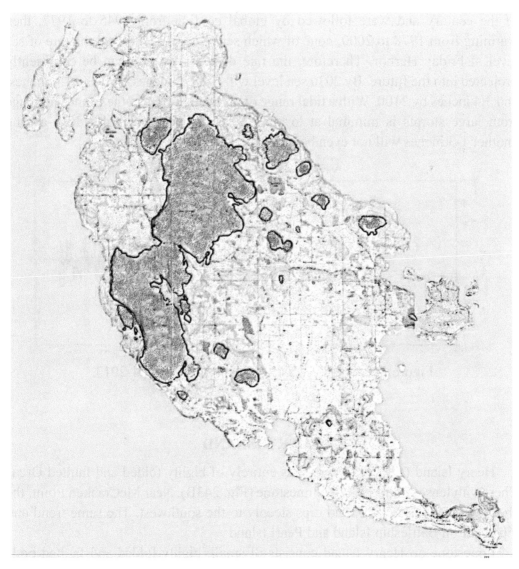

Figure 241. San Juan Island as it looked about 14,000 years ago. The brown
colored areas represent land 300 feet or more above sea level.
Blue areas were submerged.

Recent sea level rise

Sea level measurements have been recorded at Friday Harbor on San Juan Island since 1934 (Fig. 242). During that time, sea level has risen at a constant rate of 0.04 inches per year (4 inches per century). The 1930s were the warmest decade of the century and were followed by global cooling from 1945 to 1977, then warming from 1978 to 2000, none of which seems to have affected the rate of sea level at Friday Harbor. Therefore, the rate of sea level rise can be confidently projected into the future. By 2030 sea level will rise 0.6 inches; by 2050 1.4 inches; and 3.4 inches by 2100. With a tidal range of about eight feet in the islands, damage from large storms is minimal at low tide and maximum at high tides—adding another 1-3 inches will not even be noticed.

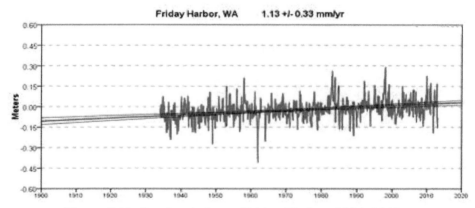

Figure 242. Sea level rise at Friday Harbor 1934-2013.

HENRY ISLAND

Henry Island (Fig. 243A) consists entirely of highly folded and faulted Orcas Chert with lenses of interbedded limestone (Fig. 243B). Near McCracken Point, the chert trends about N75°W and dips steeply to the southwest. The same trend and dip occur on Battleship Island and Pearl Island.

Limestone on Henry Island consists of small, highly folded and faulted pods and lenses interbedded with ribbon chert and volcanic rocks. Limestone beds about 80 feet long and 15 feet thick occur on a bluff overlooking Roche Harbor on the northeast coast of the island. Some of it is oolitic (composed of small spheres) and contains small shell fragments.

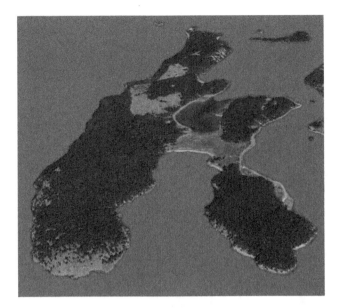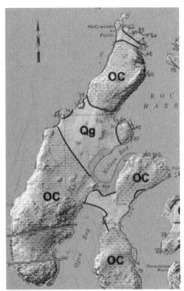

Figure 243. A. Henry Island west of Roche Harbor. B. Geologic map
OC=Orcas Chert, Qg=Ice Age sediments. Red lines are faults.

A second limestone makes a NW–SE trending ridge on the northeast coast of Henry Island opposite Pearl Island. It consists of half a dozen pods and lenses of limestone that trend about N50°W and dip steeply northward. Greywacke sandstone and ribbon chert are interbedded with the limestone.

A third limestone, dipping steeply northward, makes a 5–foot bluff above the beach along the northwest side of a small peninsula on the NW coast of the island. It is interbedded with contorted ribbon chert.

Late Ice Age shorelines occur in a down-dropped block between two faults that cross the southern peninsula (Fig. 244).

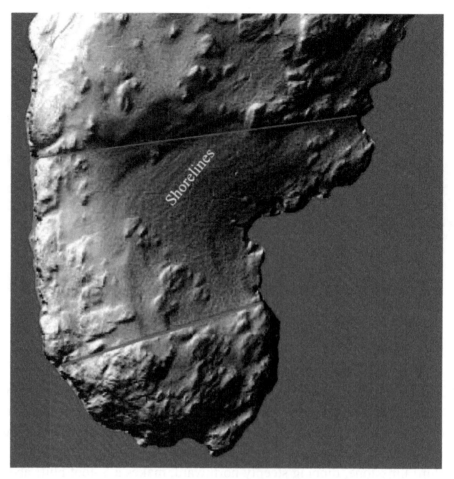

Figure 244. Southern peninsula of Henry Island where two faults have dropped down the land between the faults. Arcuate shorelines extend between the two faults.

O'NEAL ISLAND

About three-fourths of O'Neal Island (Fig. 245) is composed of diorite and amphibolite of the Turtleback Crystalline Complex. In the southern fourth of the island, it is overlain by argillite, greywacke sandstone, chert and limestone (Fig. 246). The sedimentary rocks lie on the diorite and amphibolite with a depositional contact on the SW side of the island.

171

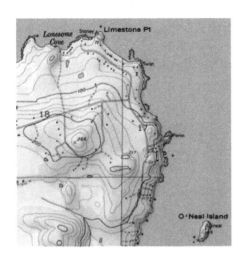
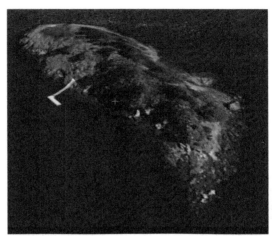

Figure 245. O'Neal Island

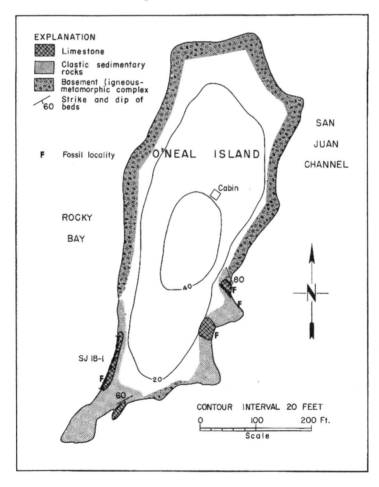

Figure 246. Geologic map of O'Neal Island. Pink=Turtleback Crystalline Complex; green=argillite, greywacke sandstone, and chert; blue=limestone. (Modified from Danner, 1966)

BROWN ISLAND

Brown Island (Fig. 247) is composed of greywacke sandstone of the Constitution Formation intruded by dikes and sills. In some places, beds have been overturned.

Figure 247. Brown Island, Friday Harbor.

TURN ISLAND

Turn Island (Fig. 248) is made up mostly of igneous intrusions.

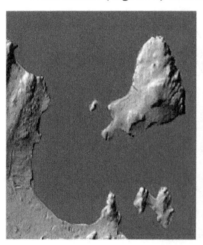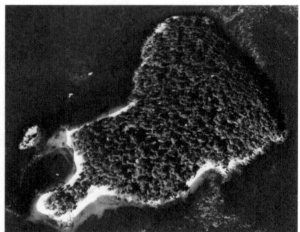

Figure 248. Turn Island.

DINNER ISLAND

Dinner Island (Fig. 249) lies offshore from eastern San Juan Island south of Friday Harbor. It is composed of Constitution thin–bedded greywacke sandstone trending northwesterly and dipping to the southwest.

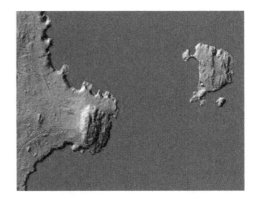

Figure 249. Dinner Island

GOOSE ISLAND

Goose Island (Fig. 250) off the east end of Cattle Point (Fig. 251) is composed of greywacke sandstone trending N55°W and dipping 35–65°NE.

Figure 250. Goose Island east of Cattle Point, San Juan Island.

SHAW ISLAND

Shaw Island (Figs. 251, 252) consists mostly of highly deformed greywacke sandstone of the Constitution Formation which has been so intensely folded and faulted that little continuity exists from one outcrop to another. Rocks along the southwest shore trend generally northwesterly and dip northeastward. Rocks along Wasp Passage trend northeasterly and dip southeastward. Beds of Orcas chert containing small limestone lenses have been intricately folded and faulted with basal greywacke sandstone of the Constitution Formation. More than half a dozen east–west high–angle faults cross the island within the Constitution Formation (Figs. 253).

174

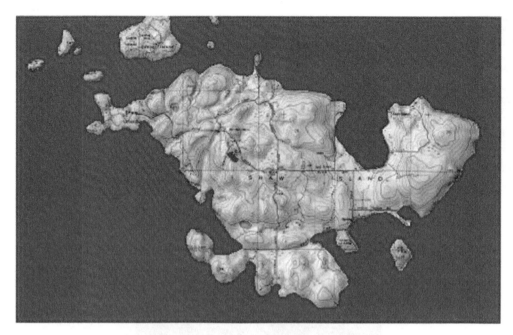

Figure 251. Topographic map of Shaw Island

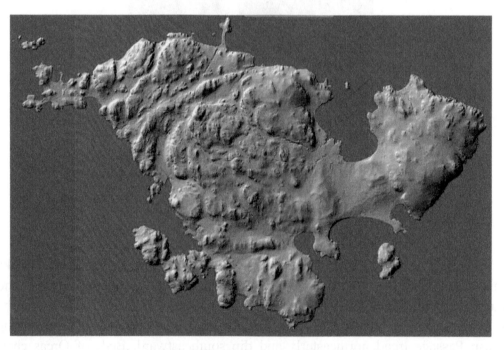

Figure 252. LIDAR image of Shaw Island.

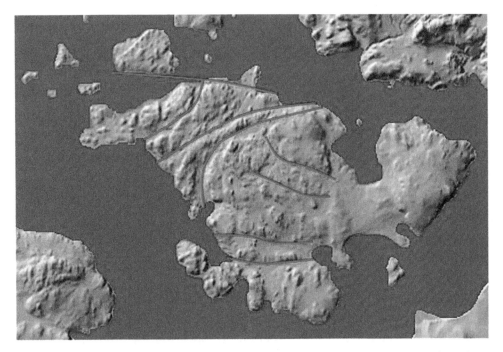

Figure 253. LIDAR map of Shaw Island. Green color represents rocks of the Constitution Formation. Red lines are faults.

Nonglacial sediments, deposited just prior to the last Ice Age, occur in sea cliffs at Indian Cove where wood in stratified sand, silt, and clay has been radiocarbon dated at 22,900 ± 200 ^{14}C years ago. These nonglacial sediments are overlain by advance outwash and glacial till of the Vashon Glaciation, Everson glaciomarine pebbly silt, and beach deposits of marine terraces. Beach sand and gravel containing marine shell fossils have been radiocarbon dated at 12,320 ± 90, 12,600 ± 380, and 12,620 ± 130 14C years old (about 14,000 calendar years).

CANOE ISLAND

Rocks on Canoe Island (Fig. 254) consist of greywacke sandstone and argillite of the Constitution Formation.

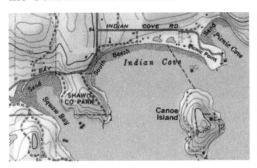

Figure 254. .Canoe Island

176

LOPEZ ISLAND

Most of Lopez Island (Fig. 255) is covered by a mantle of Ice Age glacial deposits and marine terrace deposits, but good exposures of greywacke sandstone (Fig. 256) and conglomerate with interbeds of argillite (Fig. 257), pillow basalt, (Figs. 258–61) chert, and greenstone (low grade metamorphism of basalt flows) occur at the southern end of the island. The marine greywacke sandstone and argillite commonly have graded bedding and most probably formed by submarine turbidity flows. Thin beds of ribbon chert are interbedded with argillite. All of these rocks have been strongly sheared and partially recrystallized by high pressure, low temperature metamorphism.

Pillow basalt up to 500 feet thick has been sheared and recrystallized under high pressure and low temperature to form greenschist. The green color is imparted to the rocks by crystallization of green chlorite, epidote, and actinolite.

Unaltered, poorly sorted, greywacke sandstone and conglomerate containing plant fossils were probably deposited by mudflows and rapid sedimentation into an offshore basin. They have not experienced shearing or high pressure, low temperature metamorphism like the other sedimentary rocks described above.

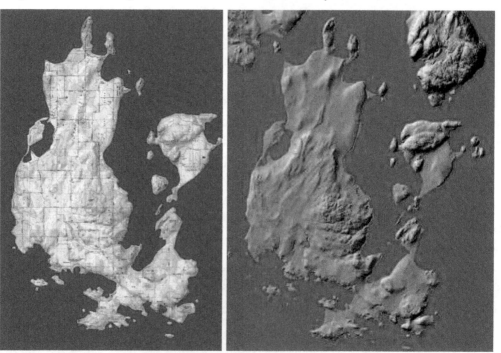

Figure 255. Topographic and LIDAR maps of Lopez Island.

Figure 256. NE–dipping sandstone beds, southern Lopez Island.

Figure 257. Black argillite cut by numerous quartz veins on southern Lopez Island.

Figure 258. Pillow lava, southern Lopez Island erupted below sea level as bulbous masses about the size of pillows.

Figure 259. Pillow lava, southern Lopez Island.

Figure 260. Pillow lava (green) overlying argillite (black) southern Lopez Island.

The age of the rocks ranges from late Jurassic to Cretaceous based on a belemnite fossil found at the base of one of the turbidity current deposits on the north side of Otter Cove. Marine, radiolarian–rich volcanic sediments are interbedded with pillow basalt at Richardson. Cretaceous microfossils and isotope dating of 124.43 ± 0.72 million years indicates a late Cretaceous age (112–115 million years).

Geologic structure

The geologic structures in southern Lopez Island are very complex. The rocks have been intensely deformed, sheared, metamorphosed under high pressure and low heat, and invaded by numerous quartz veins. The rocks were later fractured and cut by numerous, high–angle faults in a zone about a mile and a half wide, making bold, straight escarpments east–west across the island and offshore on the sea floor. This area has been called the Lopez Structural Complex (Fig. 261) and the faults here were believed to be thrust faults with large displacements. However, recent LIDAR and sonar imagery shows that the faults are really high–angle faults (Figs. 262-265), not thrust faults.

At Point Davis, argillite, greywacke sandstone, and conglomerate trend N40°W and dip 35°NE. Along the shore of Cattle Point Narrows opposite Deadman Island, the rocks trend N35°W and dip 35°NE. About a mile north of Point Davis, the rocks trend N10°E, and dip 35°SE. On Shark Reef, the rocks trend N–S and dip 15-20°E. The bend of the beds north of Point Davis is quite evident looking from the west.

Upright Head, the rocky peninsula that extends northward on northern Lopez Island is composed mostly of conglomerate with minor amounts of greywacke sandstone trending N20–40°W and dipping 35°NE. Humphrey Head, the other nearby peninsula, is composed of massive greywacke trending N10–40°W and dipping 45°NE.

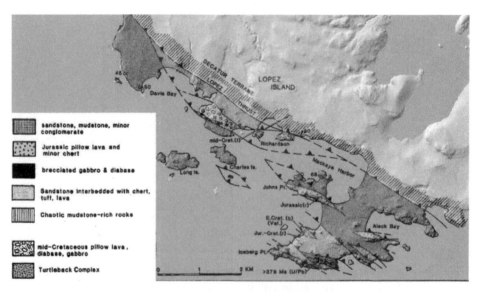

Figure 261. Geologic map of southern Lopez Island. Areas shown as thrust faults are known from LIDAR imagery to be high–angle faults, not thrust faults. (Modified from Brandon, 1989)

Figure 262. High–angle faults, southern Lopez Island.

181

Figure 263. High–angle fault scarp at the south end of Lopez Island. The vertical cliff is the topographic expression of a vertical fault.

Figure 264. Vertical fault zone, southern Lopez Island

Figure 265. Eroded fault zone between black argillite (left) and pillow lava (right), southern Lopez Island.

Watmough Bay

Black argillite and greywacke sandstone containing belemnite fossils have been tilted to vertical on the cliff at the north side of Otter Cove. At Watmough Head, (Fig. 266), steeply dipping, black, marine argillite and about 500 feet of slightly altered pillow basalt are exposed along a high-angle fault scarp (Fig. 266) that extends westward through McCardle Bay, Alec Bay, and across southern Lopez near Richardson.

The small rocky points on Watmough Head are typical roche moutonnees with glacially polished north sides, and plucked and quarried south sides. Boulder Island to the east is a rock knob with a glacially polished and grooved north end and a ragged, glacially quarried south end. Boulder Island is named for granitic erratics scattered on its surface. Many granitic erratics and other rocks types not found in the islands are scattered on beaches and upland surfaces.

Sedimentary beds near Watmough Bight trend N55°E and dip 70°NW. At Watmough Head, the trend is nearly east-west. The beach at the head of Watmough Bay is a mid–bay bar that has been built by longshore currents from glacially transported rocks. Behind the beach is a fresh–water lake.

Greywacke sandstone and thin-bedded argillite trend N 55° W and dip of 60° NE near the southeast corner of Hunter Bay. Thin-bedded greywacke sandstone trends N52°W and dips 42°N at the southeast side of Mud Bay. Shark Reef is composed of greywacke sandstone, argillite, and minor chert–pebble conglomerate.

Figure 266. Fault scarp (cliff at right) at Watmough Bay, Lopez Island.

Iceberg Point

Iceberg Point (Fig. 267) is part of the San Juan Islands National Monument. Public access is via Mackaye Harbor Road to Agate Beach County Park parking lot then by trail for about half a mile through the forest to the point.

At Iceberg Point, greywacke sandstone and interbedded argillite (Figs. 267, 268) trend about N65°W and dip 40–70° NE. They have been sheared along closely spaced planes visible in the rocks. Clusters of numerous white, quartz veins cut the argillite and sandstone (Fig 268).

Figure 267. Iceberg Point on the southern tip of Lopez Island.

Figure 268. Greywacke and argillite cut by numerous quartz veins, Iceberg Point.

Figure 269. North-dipping greywacke and argillite at Iceberg Point.

The peninsula that makes up Iceberg Point is cut by numerous high–angle faults (Figs. 270, 271). The trail to Iceberg Point from Agate Beach Park crosses a fault that extends from the easternmost point westward across the peninsula and into Outer Bay (Figs. 270). Aleck Bay is bounded by high–angle faults and the San Juan–Lopez fault makes a scarp several hundred feet high just offshore to the south.

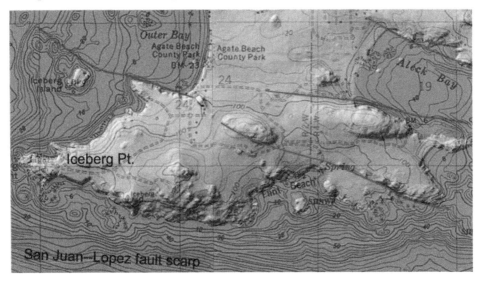

Figure 270. Faults (red) near Iceberg Point.

Figure 271. Iceberg Point. Red lines are high–angle faults. The San Juan–Lopez fault scarp lies submerged just offshore.

The Ice Age on Lopez Island

Most of the interior of Lopez Island is mantled with glacial deposits and deep glacial grooves have been scoured into the bedrock by the Cordilleran Ice Sheet (Figs. 272–274), which here was at least 6,000 feet thick. The surface of the bedrock has been polished and striated by the overriding ice. East of Iceberg Point, granite erratic boulders, carried from Canada by the Cordilleran Ice Sheet, are strewn over the landscape (Figs. 275–277).

186

Figure 272. Glacial grooves and ice-streamlined rock off Iceberg Point.

Figure 273. Grooves, striations, and polished rock made by the
Cordilleran Ice Sheet as it rode over Iceberg Point.

Figure 274. Glacial sculpted grooves, striations, and polished rock
made by the Cordilleran Ice Sheet at Iceberg Point

Figure 275. Glacial deposits and granite erratics lying on glacially sculpted, grooved, and polished rock at Iceberg Point.

Figure 276. John Whetten, pioneer of geologic research in the San Juan Islands, sitting on a granite erratic, Iceberg Point.

Figure 277. Granite glacial erratic left by the Cordilleran Ice Sheet at Iceberg Point.

Glaciomarine pebbly silt (Fig. 278) mantles Lopez Island below elevations of about 300 feet. It is overlain by raised beaches and marine shoreline deposits (Fig. 279). A radiocarbon date of 13,070 ± 130 ^{14}C years (about 15,500 calendar years) has been obtained from shells in the glaciomarine sediments and shells in the overlying raised shoreline deposits have been radiocarbon dated at 12,740 ± 150 and 12,880 ± 200 ^{14}C years (about 15,000 calendar years).

Figure 278. Late Ice Age glaciomarine pebbly silt, southern Lopez Island.

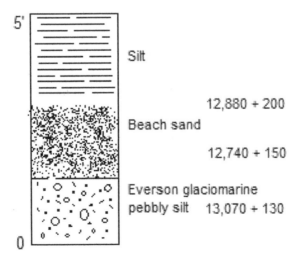

Figure 279. Ice Age deposits at Davis Bay. Numbers are radiocarbon dates in ^{14}C years before present.

Raised shorelines

During deposition of Everson glaciomarine sediments, sea level in the San Juan Islands was at least 300 feet higher than present. Following melting of the floating glaciomarine ice, sea level began to drop, leaving beach ridges of shell–bearing sand and marine terraces at successively lower levels until the present sea level was reached. The beach ridges are not easily observable from the ground or from the air, but show up in great detail on LIDAR images (Figs. 280–282). Raised former spits make sinuous ridges of sand that occur along many ridge crests (Fig. 281).

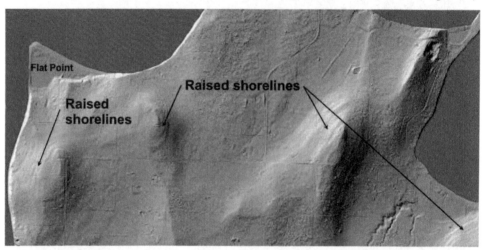

Figure 280. Raised shorelines, northern Lopez Island.

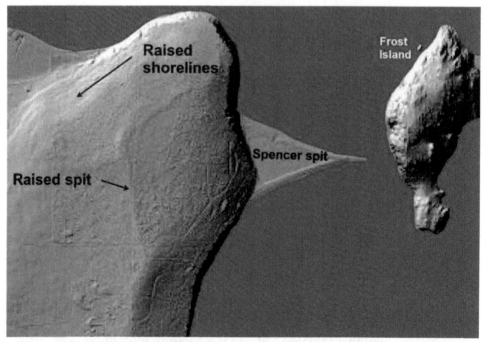

Figure 281. Raised shorelines and former spits on northern Lopez Island.

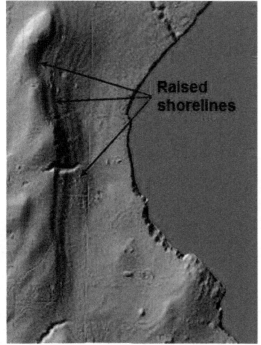 

Figure 282. Raised shorelines, eastern Lopez Island south of Spencer Spit.

Figure 283. Reconstruction of Lopez Island during the 300 foot sea level stand about 14,000 years ago. Brown= land above sea level, blue=submerged.

At the time of the high, late Ice Age, sea level stand (300 feet), Lopez Island looked quite different than it does now. Only Lopez Hill and a few small areas stood above sea level (Fig. 283). As sea level dropped to the present level, more and more of the island emerged.

Modern shoreline features

Modern shoreline processes are affecting the shape of Lopez Island. Northerly longshore transport of sand has produced a spit at Fisherman Bay that has grown almost all the way across the bay and nearly sealed it off (Fig. 284). As the spit continues to grow northward, it will eventually close off Fisherman Bay from the sea. Flat Point on the NW corner of Lopez Island is building seaward and has enclosed a lagoon and swamp (Fig. 285).

As spits build seaward, the bending of waves around islands affects longshore transport of sediment and causes spits to be built out toward the islands. When they finally reach the island, they attach it to the mainland to form a landform known as a tombolo. Spencer spit is a good example of this as it builds toward Frost Island (Fig. 286). Humphrey Head (Fig. 287) was once an island but a spit built northward and attached it to the main island.

Figure 284. A spit at Fisherman Bay has grown almost all the way across the bay and nearly sealed it off.

Figure 285. Flat Point on the NW corner of Lopez Island. Seaward extension of converging spits has enclosed a lagoon and swamp.

Figure 286. Spencer spit building seaward toward Frost Island, NW Lopez Island.

Figure 287. Humphrey Head, a tombolo made by attaching of a former island to the main island by longshore transport of sediment.

194

COLVILLE ISLAND

Colville Island (Fig. 288) is composed of thin–bedded argillite and greywacke sandstone with some conglomerate beds trending E-W and dipping 55° N.

Figure 288. Colville and Castle Islands off the south coast of Lopez Island.

CASTLE ISLAND

Castle Island (Figs. 288, 289) is composed of slightly metamorphosed, intrusive igneous rocks.

Figure 289. Castle Island. The bluff on the right is a vertical fault scarp on Lopez.

195

CHARLES ISLAND

Charles Island, off the southern coast of Lopez Island (Fig. 290), is composed mostly of thin–bedded argillite with interbeds of conglomerate and greywacke sandstone trending N70-80° W and dipping 60-75° NE. The small rocky islands northeast of Charles Island are composed of argillite with similar trends and dips.

Figure 290. Charles Island.   Figure 291. Long Island.

LONG ISLAND

Long Island, off the southern coast of Lopez Island (Fig. 291), is composed of conglomerate, massive greywacke sandstone, and thin–bedded argillite. At the west end of the island, the beds trend N45° W and dip 35° NE. On the central and eastern parts of the island, beds trend N75° W and dip 65° NE. The small islands tied to the southern end of the central part of Long Island by sand bars are composed mostly of thin–bedded argillite and greywacke sandstone.

DEADMAN ISLAND

Deadman Island is composed of greywacke sandstone trending N65–70°W and dipping 60°NE.

WHALE ROCKS

Whale Rocks are composed of greywacke sandstone trending N40–45°E and dipping 30°SE.

ICEBERG ISLAND

Iceberg Island is composed of massive greywacke sandstone cut by igneous intrusions.

ALECK ROCKS

Argillite composing Aleck Rocks trends about N75°W and dips steeply NE.

BUCK ISLAND

Buck Island is composed of greywacke sandstone cut by igneous intrusions.

BOULDER ISLAND

Boulder Island (Fig. 292) is composed almost entirely of igneous intrusions with a thin remnant of steeply north–dipping beds at its southern margin.

Figure 292. Boulder Island.

MUMMY ROCKS

Mummy Rocks are composed of greywacke sandstone trending N 60°W and dipping 45°NE.

SECAR ROCK

Secar Rock is composed of greywacke sandstone cut by igneous intrusions.

HALL ISLAND

Hall Island is composed of thin-bedded greywacke sandstone and argillite trending N70° W and dipping 50–70° NE.

CRAB ROCKS

Sedimentary beds composing Crab Rooks trend NE and dip SE.

Figure 293. Center, Ram, and Rim Islands.

CENTER ISLAND

Center Island (Figs. 293, 294) consists of thin–bedded argillite and greywacke sandstone trending N18° E and dipping about 25° SE at the southern end of the island. At the northern part of the island the trend is about N25–30° W and the dip is 30–40° NE.

Figure 294. LIDAR map of Center, Ram, and Rim Islands.

RAM ISLANDS

Ram Islands (Figs. 293–295) consist of an asymmetric ridge of tilted resistant sandstone intruded by igneous rocks. The sandstone trends parallel to the long direction of the islands and dips about 20° SE (to the left on the figure).

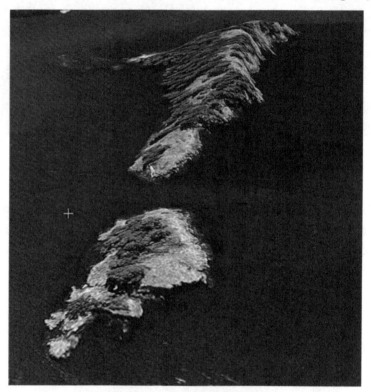

Figure 295. Ram Islands.

TRUMP ISLAND

Trump Island (Fig. 296) is composed chiefly of thin–bedded argillite and greywacke sandstone. The higher area along the east side of the island is composed of massive greywacke sandstone trending N 20 °W and dipping 45° NE.

Figure 296. Trump Island.

DECATUR ISLAND

Decatur Island (Figs. 297, 298) is composed of late Jurassic to early Cretaceous marine greywacke sandstone and argillite (Fig. 299). Decatur Head consists of conglomerate with interbeds of argillite and greywacke sandstone trending N 80-85° E and dipping about 45° NW. Conglomerate at the NW corner of the island trends N 55° E and dips 70° SE. Greywacke sandstone and argillite near Fauntleroy Point trend N75°E and dip 75° NW. Northern Decatur Island is made up mostly of greywacke sandstone, conglomerate, and argillite, cut by igneous intrusions. Thin–bedded argillite on the west side of the southernmost peninsula has been folded into an anticline. Massive greywacke sandstone and argillite making up the peninsula north of Trump Island trend about N5°W and dip steeply SW.

Fission–track dating of zircon in rocks at Decatur Head gave ages of 78.7 ± 12.4 and 61.0 ± 8.2 million years. Radiolarian microfossils from Trump Island indicate an age of late Jurassic to early Cretaceous.

Figure 297. Topographic map of Decatur Island.

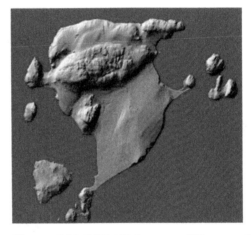

Figure 298 LIDAR image of Decatur Island.

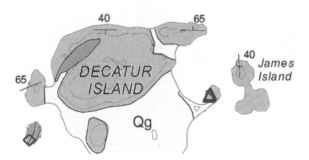

Figure 299 Geologic map of northern Decatur Island. Purple=late Jurassic to early Cretaceous marine greywacke and argillite, Qg=Ice Age sediments. Numbers are angle of dip of the beds. (Modified from Lappen, 2000)

Ice Age Deposits

Good exposures of Ice Age deposits occur at White Cliff along the east coast of Decatur Island. There, highly deformed silt and clay near the base of the bluffs are overlain by horizontally stratified sand, capped by Everson glaciomarine pebbly silt (Fig. 104). Erratic boulders dropped by the Cordilleran Ice Sheet are common occurrences. One of the largest erratics in the San Juan Islands occurs just offshore west of the island (Fig. 100). Most of southern Decatur Island is mantled with late Ice Age glaciomarine sediments and raised beach deposits (Fig. 299).

Raised shorelines

Like most of the other San Juan Islands, Decatur was submerged to elevations now 300 feet above present sea level and all but the central dome of the island was covered by the sea 14,000 years ago. As sea level dropped, multiple shorelines were etched into the topography at successively lower elevations (Fig. 300). Beach ridges and marine terraces now encircle the higher areas of the island. Figure 301 is a reconstruction of what the area would have looked like about 14,000 years ago.

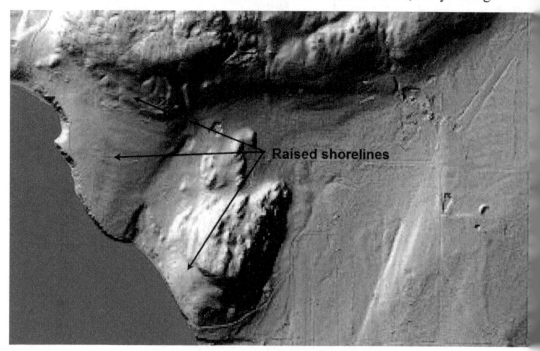

Figure 300. Raised shorelines, Decatur Island.

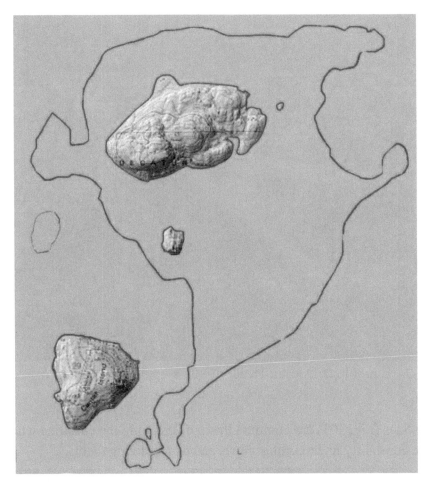

Figure 301. Reconstructed sea level around Decatur Island 14,000 years ago. Dark blue line is the outline of the present island.

Modern shoreline features

Beach sediments are continuously moved by wave action, both up and down the beach and also laterally, parallel to the shoreline. Consequently, large volumes of sediment are moved every year on all beaches. Where incoming waves approach a shoreline at an angle, sediment is driven laterally along the shoreline. If the beach makes a sharp turn, as around a point of land, the beach sediment may not be driven around the corner but continues straight ahead and is dumped in deeper water, building a spit seaward.

Islands can affect waves as they come ashore. Waves bend around islands and as a result, the shoreline behind an island can be strongly affected. The bending of the waves causes them to push sediment to the lee of an island and promote growth of a spit toward the island until it becomes attached to the mainland. Such features are known as tombolos (Fig. 302).

Figure 302. Tombolo at Decatur Head.

Landslides

Landslides (Fig. 303) are common in sea cliffs along the coastline where waves are constantly undercutting slopes and destabilizing them.

Landslide

Figure 303. Large landslide on the east coast of Decatur Island. The bluffs at the back of the landslide mark the slide plane where the mass broke away from the land above it.

JAMES ISLAND

James Island (Fig. 304) is composed of upper Jurassic to lower Cretaceous greywacke sandstone, conglomerate, and argillite, which trend about N10°E and dip to the east. They have been intruded by igneous dikes.

Figure 304. James Island

FLOWER ISLAND

Flower Island is composed of massive greywacke sandstone and conglomerate trending generally N-S and dipping 60°E.

FIDALGO OPHIOLITE, BLAKELY AND CYPRESS ISLANDS

Ophiolites are segments of the Earth's oceanic crust and underlying mantle pushed onto continental crustal rocks and uplifted above sea level. They are made up of assemblages of mostly green rocks—gabbro, serpentine, pillow lava, chert, and spilites (fine-grained igneous rocks resulting from alteration of oceanic basalt). Ophiolites are interpreted to represent the formation of new crust by spreading of crustal rocks at oceanic ridges.

Spilites are formed when basaltic lava reacts with seawater or from hydrothermal alteration when seawater circulates through hot volcanic rocks. They are composed of green chlorite, epidote, and actinolite. Hot fluids circulating through oceanic crust causes serpentinization, alteration of peridotite, gabbro, and basalt to lower temperature assemblages. Pyroxene alters to chlorite, and olivine alters to serpentine.

Fidalgo ophiolite intrusive igneous rocks include layered gabbro, diorite, granite, and basalt, which have undergone metamorphism to greenschist by crystallization of actinolite, feldspar, chlorite, and epidote.

Three isotope dates from zircon range from 170±3 to 160 ±4 million years. Radiolarian microfossils in the overlying sedimentary rocks on Fidalgo, Guemes, Lummi, Decatur, Blakely Islands and nearby smaller islands indicate a Jurassic age. Fidalgo ophiolites include serpentine and peridotite (80% olivine 20% pyroxene) with smaller amounts of dunite. Some olivine grains reach up to three inches in diameter. Fidalgo ophiolite occurs on Blakely, Cypress, Fidalgo, Hat, and Saddlebag Islands. Olivine in peridotite and dunite has commonly been replaced by serpentine.

BLAKELEY ISLAND

Almost all of the rocks on Blakely Island (Fig. 305) are igneous intrusive rocks of the Fidalgo ophiolite, including layered gabbro, diorite, granite, and basalt, which have undergone greenschist metamorphism to form actinolite, albite, chlorite, and epidote that give the rocks a green appearance.

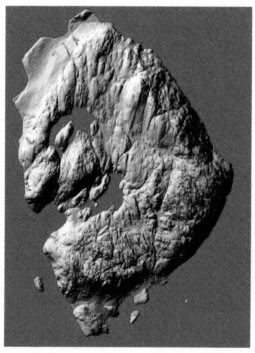

Figure 305. LIDAR image of Blakely Island.

Figure 306. LIDAR–topographic overlay of Blakely Island. Red lines are faults.

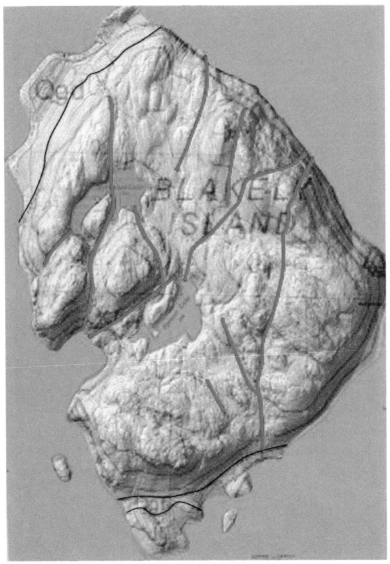

Figure 307. Geologic map of Blakely Island. Green = Fidalgo igneous intrusive ophiolite. Red lines are faults. (Modified from Lappen, 2000)

Several generally north–trending faults cutting the Fidalgo ophiolite are apparent on the LIDAR imagery (Figs. 306, 307). The northern extremity of Blakely Island is composed mostly of late Jurassic–lower Cretaceous conglomerate, trending N60°E and dipping ~60° SE.

Raised shorelines

Multiple raised shorelines are apparent on LIDAR images on the NW and southern coasts of Blakely Island. The highest shorelines are about 300 feet above present sea level. Numerous lower shorelines extend down to present sea level (Figs. 308, 309).

206

Figure 308. Raised shorelines along the NW coast of Blakely Island. The highest level is about 300 feet above present sea level.

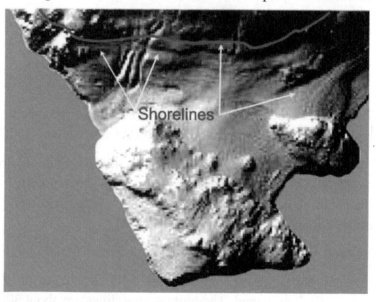

Figure 309. Raised shorelines along the southern part of Blakely Island. The highest level is about 300 feet above present sea level.

EASTERN ISLANDS

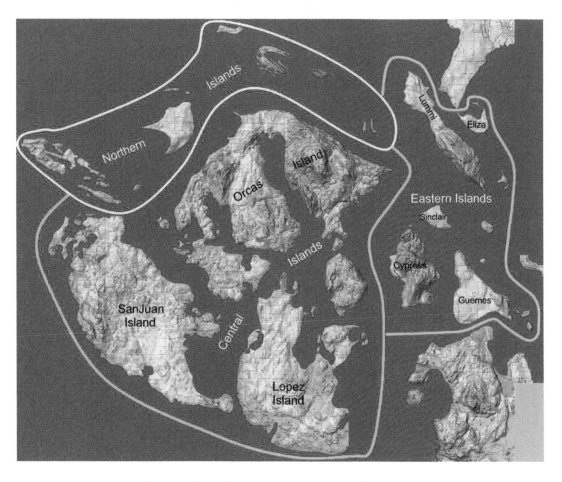

Figure 310. The eastern San Juan Islands.

CYPRESS ISLAND

Much of Cypress Island (Figs. 310, 311) has high, rugged topography, rising 1500 feet above sea level. Most of the island consists of dunite, peridotite, and serpentine (Fig. 312). These olivine–rich rocks usually have a golden, light brown color on weathered surfaces, whereas the serpentine is usually dark green. Olivine Hill at the southeast corner of the island is composed almost entirely of fresh, glassy olivine with crystals up to two inches in length.

Thin–bedded Jurassic argillite and greywacke sandstone make up the northern portion of Cypress Island. Along the shore northwest of Eagle Cliff, the beds trend N70°E and dip 30°SE. Northeast of Eagle Cliff, the beds trend N 60–75°W, and dip 25–55°SW. North of Eagle Harbor, beds trend N60°E and dip 60°SE. Igneous intrusions cut the sedimentary rocks.

Figure 311. Cypress Island.

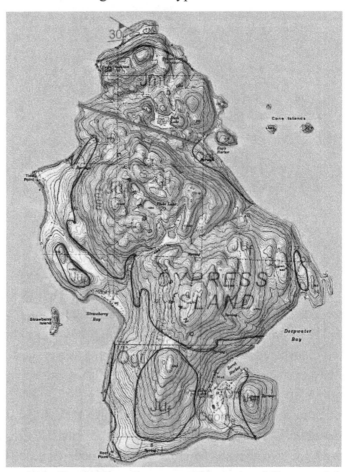

Figure 312. Geologic and topographic map of Cypress Island.
Green=dunite and peridotite, gray=Lummi Fm., yellow=Ice Age sediments.

Sea cliffs along the southern shore of the island consist of glacial till and large glacial erratic boulders are strewn over the landscape. At one point along the southern coast, glacial till has been cemented with iron oxide and converted to solid rock that rings when struck with a hammer (Fig. 313, 314).

Figure 313. Glacial tillite along the south coast of Cypress Island. The original glacial till has been so strongly cemented by iron oxide derived from nearby dunite bedrock that it is now a solid rock.

Figure 314. Till converted to solid rock (tillite) by iron cementation derived from dunite bedrock.

STRAWBERRY ISLAND

Strawberry Island consists of argillite and greywacke trending N–S with near–vertical dips.

TOWHEAD ISLAND

Towhead Island is composed chiefly of dark-colored, thin–bedded argillite, which trends N70°E and dips 40°SE.

CONE ISLANDS

The Cone Islands are composed of thin–bedded argillite trending generally N70°E and dipping 35°SE, intruded by igneous rocks. Rocks of the northern Cone Islands trend N55–70° E and dip 35–65° SE.

ALLAN AND BURROWS ISLANDS

Allan and Burrows Islands (Figs. 315, 316), just west of Anacortes, are composed of (1) large irregular masses of very coarse–grained dunite (almost entirely olivine), which typically weathers to a rusty yellow–tan color, 2) thin, irregular fine–grained dunite that has been injected into the coarse–grained dunite, and (3) thin stringers of serpentine Small amounts of black chromite are also present.

The coarse–grained dunite contain crystals of olivine half an inch in diameter. Individual crystals on rock surfaces have been differentially etched by erosion and the stand out in relief. The dunite is everywhere cut by thin stringers of serpentine.

Figure 315. Allan and Burrows Islands.

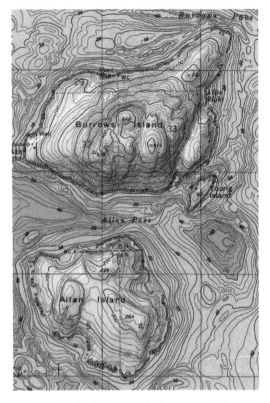

Figure 316. Allan and Burrows Islands.

GUEMES ISLAND

Most of Guemes Island (Fig. 317) is relative flat, underlain by a thick blanket of Ice Age deposits. Fidalgo igneous intrusive rocks make up two hills at the southeast corner of the island and a thin sliver of thin–bedded argillite and greywacke sandstone trend about N60°W to N75°E, and dip 70°S.

Figure 317. Guemes Island.

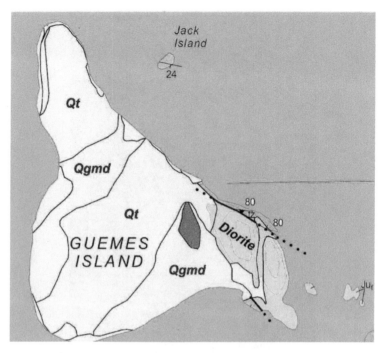

Figure 318. Geologic map of Guemes Island.
Qgmd=glaciomarine sediments,Qt=glacial till. (Modified from Lappen, 2000)

The rocks making up the two hills at the southeast end of the island consist of intrusive diorite (Figs. 318, 319) composed of altered plagioclase and hornblende. The diorite is best exposed along the east coast between Long Bay and Boat Harbor. Fine–grained granitic and dark green dikes cut the diorite. Inclusions of dark blocks of coarse–grained amphibole occur in the diorite.

Black argillite interbedded with chert occurs along the shoreline just north of Boat Harbor. The contact with the adjacent diorite intrusion is not exposed but is thought to be a fault.

Massive, metamorphosed volcanic rocks occur along the southeastern coast between Cook's Bay and Southwest Point (Fig. 319). The contact with diorite intrusive rocks to the north is not exposed but is thought to be a fault. A variety of rocks, including lava flows, volcanic breccia, metamorphosed conglomerate and sandstone, and phyllite occur along the coast between Cook's Cove and Deadman Bay.

A distinctive, green, strongly deformed, metamorphosed conglomerate and breccia occurs on the east side of Madman's Bay at Cook's Cove. The conglomerate is composed of pebbles and cobbles of chert and volcanic rocks. Almost all pebbles show pronounced flattening due to intense shearing.

A zone of intensely deformed rocks occurs along the northeast shoreline of the island separating the metamorphosed sedimentary rocks to the north from the Fidalgo igneous rocks to the south.

213

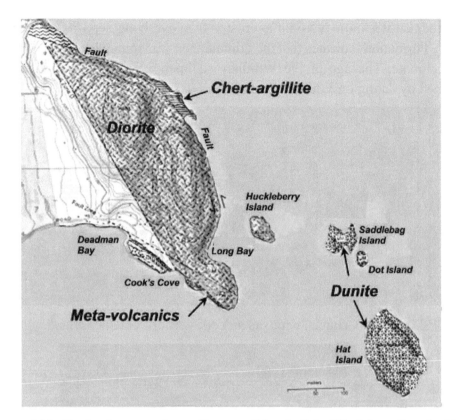

Figure 319. Geologic map of east Guemes Island. (Modified from Lamb, 2000)

Faulting

A northwest–trending fault along the northeastern shoreline (Fig. 319) appears to dip at a moderate angle to the south. A wide variety of intensely deformed rock occur along the fault zone. Many shallow, west–dipping normal faults occur along the northeast shoreline. These faults typically crosscut the earlier south–dipping faults. Offsets along these faults are generally not large (a few feet or less), but are abundant and form parallel sets. These faults are in turn cut by west–southwest and northwest trending, nearly vertical faults.

THE ICE AGE ON GUEMES ISLAND

During the last interglacial period, the Puget Lowland and San Juan Islands were a vast alluvial plain. Sand and silt were deposited on floodplains and peat was deposited in swamps. These sediments now make up the Whidbey Formation, which is well exposed in sea cliffs at Yellow Bluff on western Guemes Island (Fig. 320). The Whidbey sediments at the base of the sea cliffs consist of gray silt and sand that extend the entire length of the bluffs (Figs. 90, 91). A 2-3 foot thick peat bed occurs at the top of the exposed Whidbey (Figs. 93, 94, 321). Wood in the peat has been flattened by the weight of an overriding ice sheet and the peat has been so

214

compacted that it is more resistant to erosion than overlying deposits. Pollen in the Whidbey Formation indicates that the climate then was somewhat warmer than the present climate. The age of the Whidbey sediments is about 100,000 years as determined by dating on Whidbey Island.

Figure 320. Glacial deposits at Yellow Bluff, Guemes Island.

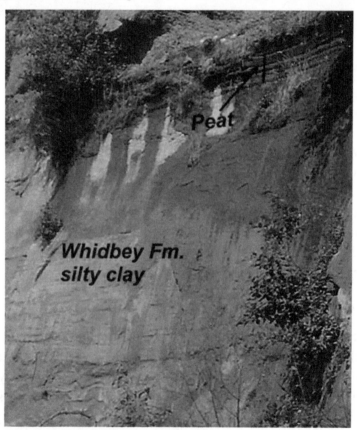

Figure 321. Whidbey Formation silty clay overlain by 100,000–year–old peat, Yellow Bluff, Guemes Island.

Glacial outwash sand and gravel overlies the Whidbey silt and sand (Figs. 322, 323, 324) and is in turn overlain by outwash sand. The age of the outwash sand and gravel is not known but may be from a period of Ice Age glaciation that occurred about 70-90,000 years ago in the lowland. The overlying outwash sand, known as the Esperance sand, is also not dated but was most likely deposited by meltwater streams in front of advancing ice of the Vashon glaciation, the last major glaciation in Washington.

In some places, Vashon till overlies the Esperance sand but is missing at Yellow Bluff. Everson glaciomarine pebbly silt caps the top of the bluffs (Figs. 324, 325). Marine shells in the Everson glaciomarine deposits have been radiocarbon at 11,700 ^{14}C years (~14,000 calendar years) at many places in the San Juan Islands and northern Puget Lowland.

Figure 322. Interglacial Whidbey Fm. silt/clay (gray bed at the base) overlain by Glacial outwash sand and gravel and Everson glaciomarine drift

Figure 323. Whidbey Fm. silt/clay overlain by glacial outwash sand and gravel and glaciomarine pebbly silt, Yellow Bluff, Guemes Island.

Figure 224. Sandy phase of the Whidbey Fm. is overlain by glacial outwash gravel and glaciomarine pebbly silt, Yellow Bluff, Guemes Island.

Figure 325. Glacial outwash gravel overlying silty clay of the Whidbey Formation, Yellow Bluff, Guemes Island.

JACK ISLAND

Jack Island (Fig. 326) is composed of deformed greywacke sandstone interbedded with chert, argillite, and phyllite and cut by numerous white quartz veins. The beds trend N 65–75°W and dip 30°SW.

At least two phases of faulting occur on the island. The oldest faults trend generally northwest with shallow to moderate southwest dips. A second set of many moderate to steep, south-dipping faults generally trend northwest and typically cut across the shallow earlier structures.

Figure 326. Jack Island.

SADDLEBAG, DOT, AND HAT ISLANDS

Saddlebag, Dot, and Hat Islands (Fig. 327) consist of serpentinized dunite and peridotite (olivine and pyroxene) with thin chromite and pyroxenite dikes. Exposures are generally a distinctive bronze to rust brown on weathered surfaces. Alternating layers of dunite and peridotite range from a few inches to several feet thick. Olivine in the dunite is fine grained and highly serpentinized (90 95%). Contacts between rock bodies are near vertical everywhere.

Figure 327. Huckleberry, Saddlebag, Dot, and Hat islands

HUCKLEBERRY ISLAND

The rocks on Huckleberry Island (Fig. 327) consist of diorite composed almost entirely of hornblende and feldspar, probably equivalent to the rocks exposed on Guemes Island to the west. As on Guemes, the diorite is locally crosscut by black, fine grained dikes which are exposed along the west and north shore. The dikes are generally steep, northeast trending, and about 6–8 inches thick.

Along the southwest shoreline, a white, talc–serpentinite breccia forms a conspicuous landmark. This outcrop exposes abundant boulder to cobble sized angular fragments of dark green to black, slickensided serpentine, suspended in a white, talc–rich matrix.

VENDOVI ISLAND

The rocks making up Vendovi Island (Fig. 328, 329) have been mapped as belonging to the Lummi Formation, which includes metamorphosed basalt, gabbro, and chert, along with minor limestone, argillite, and serpentine (Fig. 330). Microscopic marine radiolarian fossils in metamorphosed chert indicate a Jurassic age for deposition of the chert.

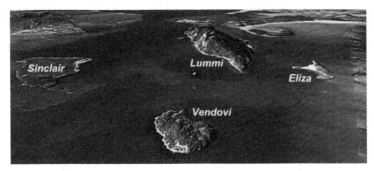

Figure 328. Sinclair, Vendovi, Lummi, and Eliza Islands.

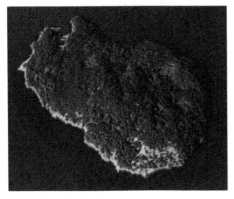

Figure 329. Vendovi Island.

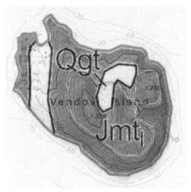

Figure 330 Geologic map of Vendovi Island. Brown=Lummi Fm. (Modified from Lappen, 2000)

ELIZA ISLAND

Eliza Island (Figs. 328, 331) is made up of three rocky points of metamorphosed greywacke sandstone and argillite of the Lummi Formation, connected by unconsolidated glacial sediments and modern beach deposits. Graded beds are very common in the sandstone and many beds are vertical and overturned.

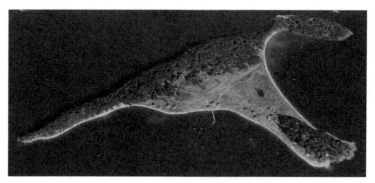

Figure 331. Eliza Island

The northern point exposes thinly interbedded metamorphosed greywacke sandstone, argillite, and phyllite (Fig. 332). The western point and southern points are predominantly medium grained metamorphosed greywacke sandstone interbedded with argillite. Folds are well exposed on both the north and south points, and most have overturned, north–dipping limbs.

At least two stages of faulting occurred in the rocks. Earlier low–angle faults are cut by steeply dipping faults that have move laterally. Sonar bathymetry suggest that Eliza Island is bounded by faults (Fig. 333).

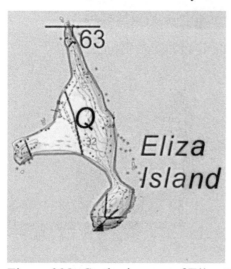

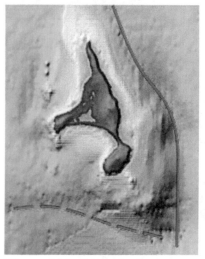

Figure 332. Geologic map of Eliza Island. (Modified from Lappen, 2000)

Figure 333. Sonar bathymetry map of the sea floor off Eliza Island. Red lines are faults (Base map by NOAA)

Isotope dating indicates and age of 137 million years (early Cretaceous) for rocks on Eliza Island. This is interpreted to date the time of metamorphism, rather than the depositional age.

SINCLAIR ISLAND

The southern hill on Sinclair Island (Figs. 328, 334) is composed of metamorphosed conglomerate, greywacke sandstone, and argillite of the Lummi Formation (Fig. 335). The rocks exhibit graded bedding, ripple marks, and flame structures. Microscopic marine radiolarian fossils in metamorphosed chert indicate a Jurassic age for deposition of the chert. Most of the island is underlain by Ice Age glaciomarine sediments.

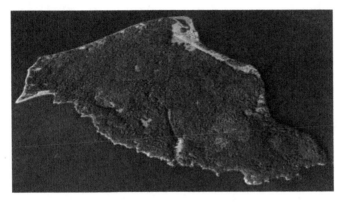

Figure 334. Sinclair Island.

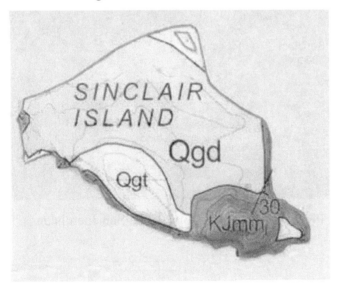

Figure 335. Geologic map of Sinclair Island. Brown=Lummi Fm., Yellow=Vashon till, green=glaciomarine sediments (Modified from Lappen, 2000)

NORTHERN ISLANDS

The northern San Juan Islands consist of sandstone, shale, and conglomerate of Cretaceous, Eocene, and Jurassic age. Unlike the older rocks to the south, the sedimentary rocks in these islands are unmetamorphosed, have not been intruded by igneous rocks, do not have extensive quartz veins, and although they have been folded and faulted, they have not been intensely internally sheared.

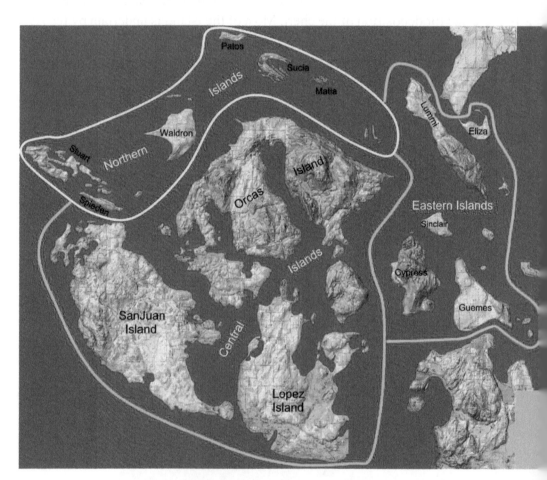

Figure 336. Subdivisions of the San Juan Islands.

SUCIA ISLAND

Figure 337. Sucia Island at sunset.

Sucia Island (Figs. 337-339) forms the nucleus of a group of islands, including Sucia Island, Little Sucia, Herndon, South Finger, North Finger, Ewing, Cluster, and Wiggins, islands, separated by many channels and bays, Echo, Shallow, Fossil, and snoring bays and Ewing and Fox coves. All of the island and waterways are part of a syncline, a trough-shaped fold that has been tilted to the east and eroded so the resulting form is U–shaped, open to the east. Since the folding of these rocks, the region has been leveled by erosion, and because of the inclined trough-like structure of the rocks, the outcrop of each bed has the form of a horseshoe. All of the islands are part of Sucia Island State Park.

The islands are composed of sandstone ridges that are more resistant to erosion than interbedded shale.The shale has been etched out by erosion, leaving the resistant sandstone as ridges standing above valleys, many of which are now filled with sea water to form bays and coves.

Figure 338. Sucia syncline, a trough–shaped fold tilted toward the top of the photo.

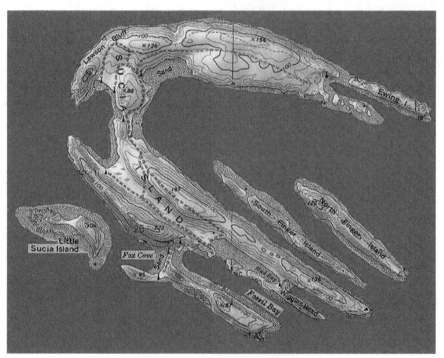

Figure 339. Sucia Island.

The rocks making up the islands consist of conglomerate, sandstone, and shale of the Nanaimo Group that were deposited in shallow seas and deltas that covered the area during the late Cretaceous Period about 60–90 million years ago and the Eocene about 50 million years ago. Over millions of years, the muddy deposits were compacted by the weight of overlying sediments, and sand grains and pebbles were cemented together by chemicals from groundwater percolating through them to form sandstone and conglomerate. The total known thickness of the Nanaimo and Chuckanut sandstones is about 3,400 feet, but this includes only the portion of the rocks above sea level. Sonar imagery indicates that it is only about half of the thickness of the stratigraphic section making up the submerged platform upon which Sucia rests.

The Namaimo Group rocks at Sucia (Fig. 340, 341) consist of three distinct units, informally named the Sucia conglomerate (Fig. 342), the Fossil Bay Formation, and the Johnson Point Formation (Easterbrook, 1958).

Time	Formation	Rock type	Thickness (feet)	Description
Eocene	Chuckanut sandstone		455	Cross-bedded sandstone and conglomerate of North Finger Island
			210	Not exposed, probably shale
			490	Cross-bedded sandstone and conglomerate of South Finger Island
			325	Not exposed, probably shale
			105	Cross-bedded sandstone and conglomerate
			290	
			330	Not exposed, probably shale
			170	Cross-bedded sandstone
			610	Not exposed, probably shale
				Black shale
Upper Cretaceous	Fossil Bay sandstone		565	Richly fossiliferous silty sandstone interbedded with 6-10" limestone beds
	Sucia conglomerate		90	Quartz-schist pebble conglomerate

Figure 340. Stratigraphic section of the Nanaimo sandstone and conglomerate and Chuckanut sandstone. (Modified from Easterbrook, 1958)

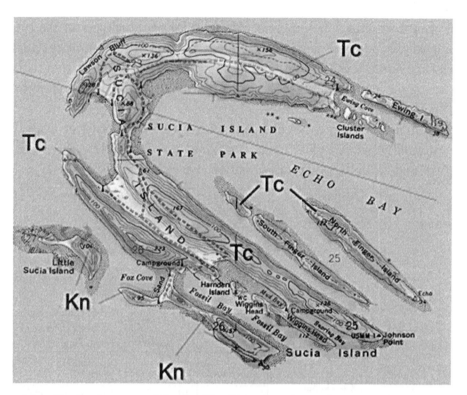

Figure 341. Geologic map of Sucia. Kn=Nanaimo Fm., Tc=Chuckanut sandstone.

Figure 342. Sucia conglomerate composed entirely of phyllite, quartz, and greenschist derived from the underlying rock. The 'swiss-cheese' appearance of the exposure is caused by honeycomb weathering. East end of Fossil Bay

CHUCKANUT FORMATION

All of Sucia, except the ridge south of Fossil Bay and Little Sucia Island, is composed of Chuckanut sandstone, shale, and conglomerate (Fig. 343), which lie directly on Fossil Bay sandstone. The sandstone is composed mostly of quartz and feldspar sand grains, which were probably derived from mechanical disintegration of granite and deposited by streams on floodplains or deltas. Pebbles in the conglomerate have been well rounded as a result of stream transport and consist of granite, andesite, argillite, quartz, and phyllite. Most pebbles average about an inch in diameter but range up to six inches. The total thickness of the beds, including rock covered by the waters of several channels, is about 3200 feet.

Both the sandstone and conglomerate are extensively cross bedded (Fig. 82, 83). Crossbedding is a type of layering of sediment deposited on the downstream side of channel bars. Most sediments are deposited with horizontal bedding, but the stratification in bars is inclined at an angle downstream on the backside of the bars.

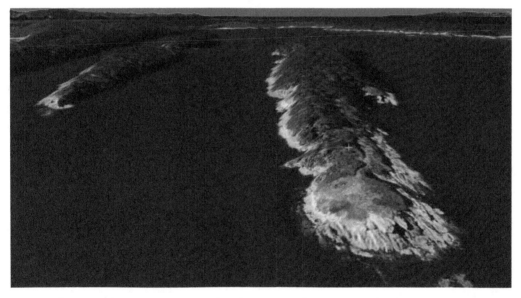

Figure 343. NW dipping Chuckanut sandstone, North Finger Island.

Geologic structure

The originally horizontal sedimentary rocks of Sucia have been folded into a southeasterly plunging syncline (Fig. 344). The structure resembles a trough that has been tilted to the southeast. The resistant sandstone beds that make up the elongate islands from Echo Bay to the ridge south of Fossil Bay all dip to the northeast whereas the ridges northeast of Echo Bay dip in the opposite direction, to the southwest.

Figure 344. Geologic cross section of Sucia Island. The beds were originally horizontal but have now been folded into a trough-like shape and tilted to the SE.

Sucia Island has long been recognized as simple synclinal fold, but sonar imagery now shows that it is actually part of a much larger, more complicated structure that makes up a largely submerged, mesa–like platform bounded by fault scarps on the north, south, and east sides (Fig. 345).

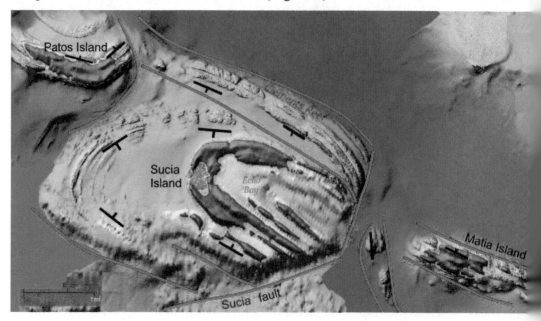

Figure 345. Geologic structure of Sucia. (Base map by NOAA)

Faulting

Sonar imagery in Figure 345 shows offsetting of the sea floor by faults between Patos, Sucia, and Matia Islands. Fault scarps are mostly several hundred feet high but a few reach 1000 feet. Evidence for the fault origin of the scarps includes abrupt truncation of beds and the length, straightness, and constant steepness of the scarps. The fault along the southern margin of Sucia truncates the beds there and extends westward for several miles, cutting across tilted beds north of Waldron Island. Just to the north, beds along the south side of the submerged platform are truncated by another fault. Beds of Nanaimo sandstone at Little Sucia Island appear to be the same as submerged, truncated beds south of the fault (Fig. 345).

The curving, outermost beds at the western apex of the submerged syncline (Fig. 345) must be Nanaimo sandstone. These beds curve around the nose of the syncline and extend eastward until ending abruptly against a fault just offshore from northernmost Sucia. The western end of the syncline ends suddenly at a north–south trending fault scarp (Fig. 345). Just east of this scarp, beds of Chuckanut sandstone on Matia dip to the north, but the rest of the syncline is missing altogether. Three sandstone ridges on Matia appear to correspond to three prominent sandstone ridges on Sucia.

Clements Reed, a long submarine ridge made of several resistant sandstone beds of Chuckanut sandstone, extends eastward from Patos Island almost to Matia Island (Fig. 345). These beds dip o the south like those on the north limb of the Sucia syncline west of Ewing Island, but are separated by a fault (Fig. 345).

A similar truncation of beds occurs on the south limb of the Sucia syncline. A steep fault scarp marking the southern margin of the Nanaimo conglomerate at the entrance to Fossil Bay extends westward but ends abruptly south of Little Sucia (Fig. 345). The end of this bed was once connected to the outer scarp that ends abruptly at the Sucia fault (Fig. 345).

Figure 346. Deformed beds along a fault near Fossil Bay.

Ice Age deposits

Like the rest of the San Juan Islands, Sucia was overridden by the Cordilleran Ice Sheet. Granite erratic boulders (Fig. 347) are strewn over the landscape and deposits of glacial till and glaciomarine sediments are preserved in low places. When the ice sheet melted and broke up, sea level was at least 300 feet higher than at present and covered with floating ice, and glaciomarine pebbly silt with cobbles and boulders were dropped on the sea floor as the floating ice melted and released enclosed sediment. Marine fossils are abundant in places, especially at the head of Fossil Bay (Fig. 348). Radiocarbon dates from the shells indicates an age of 11,700 ^{14}C years (about 14,000 calendar years).

Figure 347. Granite erratic boulder from Canada on the beach at Fossil Bay.
(Photo by George Mustoe)

Figure 348. Marine fossils radiocarbon dated at 11,700 ^{14}C years (~14,000
calendar years) in glaciomarine sediments at the head of Fossil Bay.

COASTAL LANDFORMS

The shores of Little Sucia Island, as well as the exposed shores of the other islands of the group, are fringed by flat rocky shelves which have been formed by wave erosion over thousands of years (Fig. 349, 350).

Figure 349. Wave cut bench, Little Sucia.

Figure 350. Wave–cut bench, Little Sucia.

WEATHERING FEATURES

Honeycomb weathering

A peculiar type of weathering known as honeycomb weathering is common on the surfaces of Nanaimo and Chuckanut sandstones (Figs. 351, 353). Hardening of the outer surface of sandstone by cementation also produces unusual features (Fig. 352).

Figure 351 Honeycomb weathering. Photo by George Mustoe)

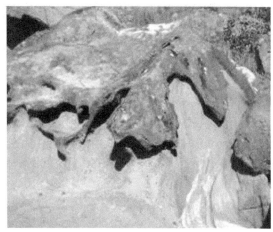

Figure 352. Weathering and erosion of sandstone. (Photos by George Mustoe)

Figure 353. Honeycomb weathering in sandstone, Sucia Island.

Indian middens

Native American Indians inhabited the San Juan Islands for thousands of years and archeological sites are common in the islands. Among the most common and the most obvious are shell middens. These are accumulations of clam shells tossed onto a shell heap by the Indians. They are recognizable as layers composed entirely of shells mixed with a little soil. A good example occurs on Ewing Island (Fig. 354).

Figure 354. Indian shell midden, Ewing Island. (Photo by George Mustoe)

MATIA ISLAND

Matia Island (Fig. 355, 356) is composed of three parallel ridges of resistant Chuckanut sandstone and conglomerate trending N68°W and dipping 68°N, separated by elongate valleys filled with glacial deposits (Fig. 357). Stream crossbedding and coalified trees are common in the sandstone. An isotope age of 49.7 ± 9.2 million years was obtained from sandstone at the southwestern corner of the island.

Figure 355. Topographic map of Matia Island.

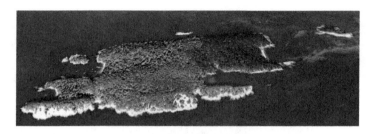

Figure 356. Matia Island.

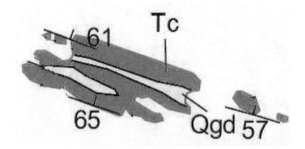

Figure 357. Geologic map of Matia Island. All beds dip northward. Numbers are angles of dip of the beds. Tc=Chuckanut sandstone, Qgd=glacial deposits.

Sandstone	250+ feet
Conglomerate and sandstone	435 feet
Shale and sandstone	332 feet
Conglomerate and sandstone	375 feet
Shale and sandstone	180 feet
Sandstone with some conglomerate	450+ feet
Total	2022+ feet

Sucia, Patos, and Matia Islands are all folded and faulted remnants of Chuckanut sandstone. Sucia is a large syncline whose axis tilts eastward, partly submerged and partly above sea level, but both Patos and Matia are made of northward tilted beds of Chuckanut sandstone (Figs. 357, 358). The three ridges of sandstone that make Matia Island appear to be the same three beds that compose similar three ridges on the southern flank of the Sucia syncline, but the Matia beds dip in only one direction, are cut off from the Sucia syncline by the Echo Bay fault, and are located north of what would be the continuation of the beds on Sucia. The northernmost beds on Sucia dip southward, just the opposite of the beds on Patos. Clements Reef, a long set of NE–dipping beds just north of Sucia dip, southward until cut off by the Echo Bay fault that crosses the east end of Sucia.

Matia Island rests on submerged platform whose northern margin is truncated by a fault (Fig. 358, 359). The southern margin of the platform is probably also a fault but the scarp is not as smooth.

Figure 358. Sonar image of the Matia Island submarine platform. (NOAA)

Figure 359. Matia fault–bounded block. (Base map by NOAA)

CLARK AND BARNES ISLANDS

The rocks on the Barnes and Clark Islands (Fig. 360, 361) are composed of Nanaimo coarse conglomerate with interbeds of sandstone and shale trending N12°E and dipping generally 80°SE. On the north end of Clark Island, the beds trend N 8°W and dip 50°SW. Beds on the southeast end of Clark Island trend N30°E, and dip 30°SE. The islands and reefs making up the Sisters group just south of Clark and Barnes Islands have similar trends and dips. Clark and Barnes Islands rest on a largely submerged platform bounded by north-south faults and cut off at the south end by an E-W fault (Figs. 362–364).

238

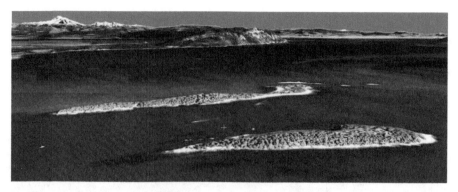

Figure 360. Clark and Barnes Islands.

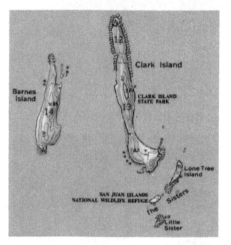

Figure 361. Clark and Barnes Islands.

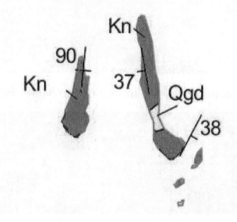

Figure 362. Geologic map of Clark and Barnes Islands. Numbers are angles of dip of the beds. Kn=Nanaimo Fm., Qgd=glacial sediments. (Modified from Lappen, 2000)

The sonar image in Figure 363 shows that Clark Island lies on the limb of an anticline, most of which is submerged below sea level. Nanaimo sandstone beds on most of Clark Island dip to the west but bend around the nose of the anticline to dip eastward at the south tip of the island (Fig. 362). The Sisters Islands just offshore to the east are made of slightly younger beds on the eastern flank of the anticline (Fig. 363). A fault truncates the limbs of the anticline on the NE flank and another fault truncates the southern part of the structure (Fig. 363, 364). The submerged beds on the east flank of the anticline are absent on the west flank, perhaps because of faulting.

Foraminifera microfossils on both Clark and Barnes Islands indicate a late Cretaceous age of the rocks. Isotope dates of 89.6 ± 12 and 105 ± 10 million years were obtained from sandstone on Barnes Island.

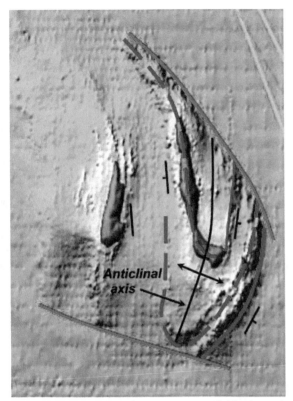

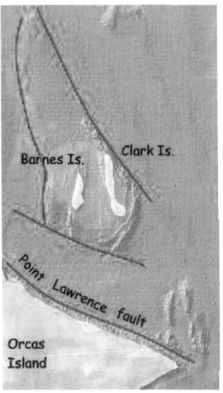

Figure 363. Geologic structure of Clark and Barnes Islands. Barnes Island is an anticline whose axis is tilted to the south. Red=faults, blue lines=submerged beds. (Base map by NOAA)

Figure 364. Faults bounding the submerged platform upon which Clark and Barnes Islands rest. (Base map by NOAA)

STUART ISLAND

The Stuart Islands (Figs. 365, 366) are the westernmost of the San Juan Islands and part of the group of islands across the northern part of the area made up of Nanaimo sandstone, shale, and conglomerate that were originally deposited in a 100-mile-long sedimentary basin extending to the northern end of Vancouver Island. The Nanaimo beds on Stuart Island have been strongly tilted (Fig. 367, 369, 370) and folded into a structural arch (anticline) and structural trough (syncline) and offset in places by faulting (Fig. 368).

The topography of the Stuart Island is directly related to the resistance to erosion of the Nanaimo beds. The upturned edges of the resistant Extension conglomerate make ridges that form the backbone of the island and the overlying and underlying, less resistant shale and sandstone have been etched out by erosion to form the main valleys and Prevost and Reid Harbors.

Figure 365. Stuart Island.

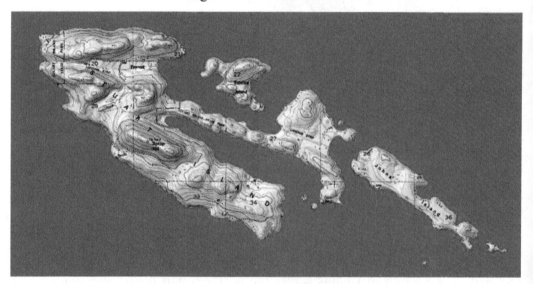

Figure 366. Stuart Island group.

The Nanaimo beds were originally deposited as sand, mud, and gravel on the floor of an ancient sea that covered the area during the Cretaceous Period about 50-90 million years ago. Shells of marine organisms living in the sea were deposited with the sediments and are now preserved in the rocks as fossils. The rocks consist of three distinct units: (1) Pender sandstone and shale, (2) Extension conglomerate, and (3) Haslam shale and sandstone.

The Haslam Formation, the lowermost unit, is about 1000 feet thick and consists mostly of shale with interbedded lenses of sandstone and conglomerate. A few fossils occur in the rocks, but are rare. Some of the carbon-rich sandstone contains scattered plant fragments. The base of the unit is not exposed. The

241

Extension conglomerate overlies the Haslam sandstone and shale. It is about 500 feet thick and consists of massive conglomerate that makes up the backbone of the Stuart Island group. Because of its resistance to erosion, it stands up as a bold ridge wherever it occurs. The Pender Formation consists of about 500 feet of sandstone and shale that overlie the Extension conglomerate. Marine fossils, mostly the genus *Inoceramus*, are abundant at Fossil Cove at the southwestern end of the main island.

Figure 367. Steeply dipping Nanaimo sandstone, northern Stuart Island.

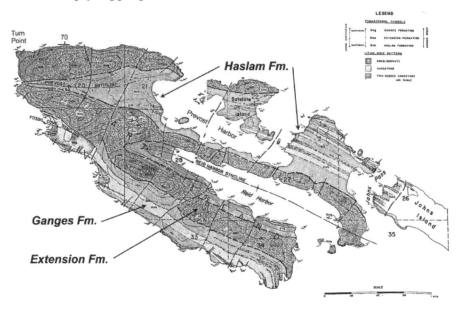

Figure 368. Geologic map of Stuart Island. (Modified from Mercier, 1977)

Figure 369. Vertically tilted beds of Nanaimo sandstone and shale,
NE Stuart Island.

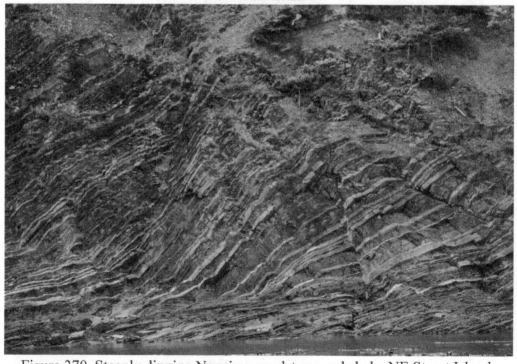

Figure 370. Steeply dipping Nanaimo sandstone and shale, NE Stuart Island.

243

Geologic Structures

Extension conglomerate on the northern part of the main island has been folded into an anticline whose axis has been tilted to the west (Fig. 371, 372). Erosion of this tilted structural arch has formed a V–shaped ridge making up the northwestern part of the island west of Prevost Harbor. Shale in the core of the anticline is less resistant to erosion than the conglomerate and has been selectively eroded to form the main valley at Prevost Harbor and the inlet to the east. Conglomerate beds along the north shore of the island dip 65–75° N and the same bed dips 65–75°S along the south side of Prevost valley.

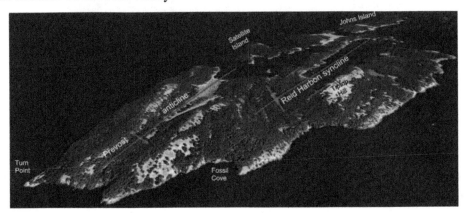

Figure 371. Geologic structures on Stuart Island. Beds dip away from the axis of the Prevost anticline and toward the axis of the Reid Harbor syncline.

Lundberg Mt. consists of a resistant conglomerate and sandstone bed dipping steeply to the north. Good views of these steeply dipping beds may be seen by boat along the north shore of the island between Charles Point and Turn Point.

South of Turn Point, the beds bend around the axis of the anticline and dip southward along the ridge that makes up Mt. Stuart. The valley between Lundberg Mt. and Mt. Stuart results from erosion of less resistant shale beds that underlie the ridge–making sandstone and conglomerate. The geologic structure becomes more complicated to the east. Whereas the axis of the anticline follows the valley between the two ridges of resistant rock, the axis of the fold does not follow the center of the elongate bay west of Prevost Harbor. The dip of the beds on the northern shore of Satellite Island is to the north, but on the south shore the beds dip to the south. Thus, the axis of the anticline leaves the bay, passes through Satellite Island, crosses the northern end of the peninsula to the east and disappears below sea level. How the thick, resistant bed making up Lundberg Mt. can disappear with no further topographic expression poses a mystery. Two possibilities are (1) the resistant bed is truncated by faulting, or (2) the composition of the bed changes laterally to a less resistant rock (shale) and has been removed by erosion.

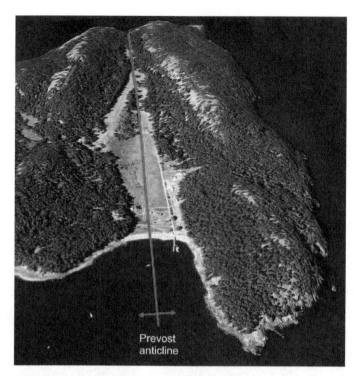

Figure 372. Anticline at Prevost Harbor looking NW. Resistant Nanaimo conglomerate dips away from the central axis in Prevost Valley. Prevost Harbor and the valley between the two ridges has been etched out by erosion of less resistant shale.

Figure 373. Reid Harbor syncline. Conglomerate beds dip towards the axis of the syncline in Reid Harbor. The same erosional and structural valley extends to the western shoreline of the island, flanked on each side by more resistant sandstone beds. On the south side of the valley this resistant bed forms a high elongate ridge called Tiptop Mountain. These same resistant beds appear again on the north side of the valley and make up the narrow arm which separates Reid Harbor from Prevost Harbor.

Beds on both sides of Reid Harbor dip toward the center of the bay, forming a syncline (Fig. 373). The resistant bed that forms Mt. Stuart comes up on the south side of the axis of the syncline to form Tiptop Mt. Beds along the southern shore of the island all dip northward into Reid Harbor syncline, but are offset along faults transverse to the ridge. The resistant bed forming Tiptop Hill (Fig. 373) is terminated abruptly by faults at both its western and eastern ends (Fig. 374). The resistant bed making Tiptop Hill does not continue eastward on the sea floor, apparently as a result of fault displacement.

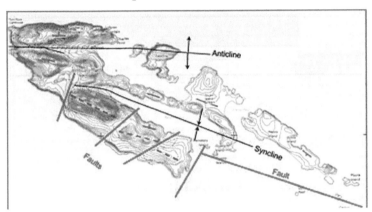

Figure 374 Folds and faults on Stuart Island.

The southern part of Stuart Island is made up of the same conglomerate folded into a structural basin whose axis trends east–west through Reid Harbor (Figs. 373, 374). Conglomerate beds along the north side of Reid Harbor dip 65–75° southward into the axis of the basin. Conglomerate making up the south side of Reid Harbor dip 50° northward into the basin. The basin is shaped like an elongate dish with Reid Harbor in the center. The conglomerate that makes the rim of the Reid Harbor basin is breached at the southeast end, allowing the sea to inundate the basin.

Faulting

The relationship of the massive Extension conglomerate to the topography is generally what would be expected of such a resistant bed, but its continuation is breached in several places—(1) the north inlet to Prevost Harbor, (2) the unnamed bay east of Satellite Island, (3) Johns Pass, (4) the break in the ridge northwest of Tiptop Hill, and (5) the east entrance to Reid Harbor. The two most common causes for such breaching of a resistant bed are local changes in composition or interruption by faulting.

Several faults extend eastward from Stuart Island on the sea floor (Fig. 375). The position of the faults is known by the direction of dip of the beds in the small islands east of Stuart Island.

246

Figure 375. Sea floor faults east of Stuart Island. (Base map by NOAA)

JOHNS ISLAND

Johns Island, an elongate, northwest–southeast trending island (Fig. 376), is a continuation of the massive, southwest–dipping conglomerate that makes up the northern coast of the main island (Fig. 377).

The stratigraphic section exposed on Johns Island is as follows:

Conglomerate and cross–bedded sandstone	320+ feet
Thick-bedded sandstone, some shale interbeds	300 feet
Conglomerate and cross–bedded sandstone	360 feet
Shale and shaly sandstone	130 feet
Conglomerate	40 feet
Rapidly alternating sandstone, shale, and conglomerate	400+ feet
Total	1550+ feet

At the east end of Johns Island the beds trend N65°W and dip about 35°SW. At the west end of the island the beds trend generally N 60°W and dip 50–55°SW.

Figure 376. Johns Island looking NW.

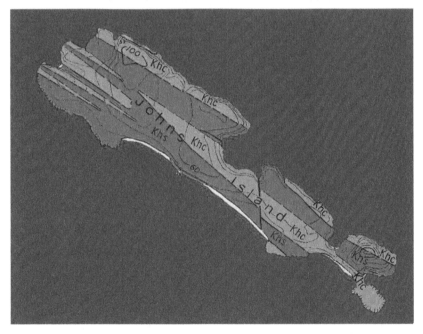

Figure 377. Geologic map of Johns Island. The beds making up Johns Island dip SW into the Reid Harbor syncline but do not reappear on the south side of the axis of the syncline as would normally be expected. Beds at Gossip Island bend around the end of the synclinal axis to form a closed basin. (Modified from Johnson, 1978)

SATELLITE ISLAND

Satellite Island (Fig. 378) is composed of Nanaimo sandstone, conglomerate, and shale, trending generally N80°W and dipping very steeply. The anticline on Stuart Island extends across Prevost Harbor to Satellite Island.

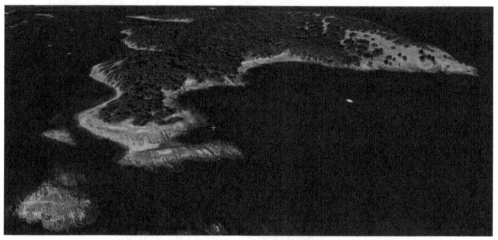

Figure 378. Satellite Island looking west.

Figure 379. Satellite Island topographic map.

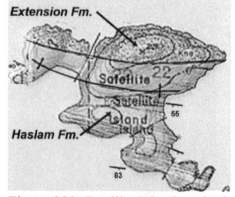

Figure 380. Satellite Island geologic map. (Modified from Johnson, 1978)

Satellite Island is a continuation of the same conglomerate bed that makes the north flank of the Prevost Harbor anticline. Cactus Islands and Gulf Reef are composed of the resistant conglomerate bed of the main island but separated from it by open water. Sandstone and conglomerate making up Cactus Islands dip to the south, are not continuous with the structures of Stuart Island, and no indications of Stuart Island rocks occur on the sea floor east of the fault that truncates the rocks south of Gossip Island.

FLATTOP ISLAND

Flattop Island (Fig. 381) is composed of about 250 feet of coarse conglomerate, which is underlain by about 35 feet of thin–bedded shale and sandstone (Fig. 382, 383). The beds trend N65°–70°E and dip 22–26°SE.

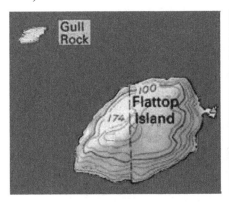 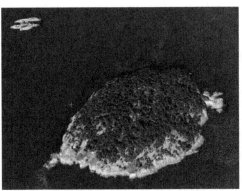

Figure 381. Flattop Island.

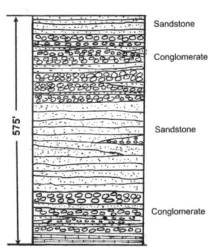 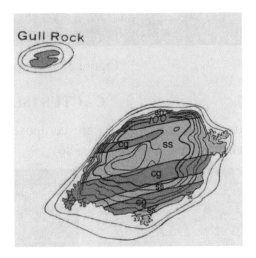

Figure 382 Stratigraphic section of Nanaimo beds on Flattop Island. (Modified from Johnson, 1978)

Figure 383. Geologic map of Nanaimo sandstone (ss) and conglomerate (cg) on Flattop Island. (Modified from Johnson, 1978)

GULL ROCK

Gull Rock is about 500 yards northwest of Flattop Island and rises only about 30 feet above sea level. It consists of two beds of resistant conglomerate, separated by more easily eroded shale. The trend of the beds is N65°E. and the dip is 65°SE.

RIPPLE ISLAND

The rocks on Ripple Island (Fig. 384) are equivalent to those across the channel on Johns Island to the west. Sandstone is overlain by about 125 feet of conglomerate and thin-bedded shale and sandstone. The beds trend N80°W and dip 45°SW.

Figure 384. Ripple Island.

CACTUS ISLANDS

Cactus Islands (Figs. 385) are composed of Nanaimo conglomerate, cross-bedded sandstone, and shale (Fig. 386)

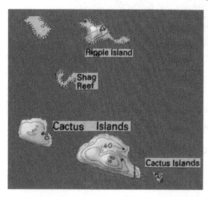 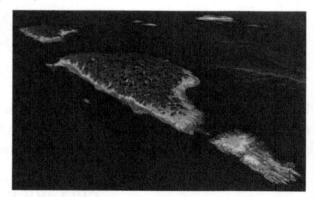

Figure 385. Cactus Islands

The rocks exposed on East Cactus Island consist of:

Massive buff-colored sandstone	170'+
Conglomerate	25'
Cross-bedded sandstone	160'
Thin-bedded shale and sandstone	40'
Cross-bedded sandstone and conglomerate	300'+
Total	695+

At the east end of Cactus Islands the average trend of the rocks is N73°W and the dip is 55°–63°SW. At the west end of the group the average trend is about N65°W and the dip is 60–66°SW.

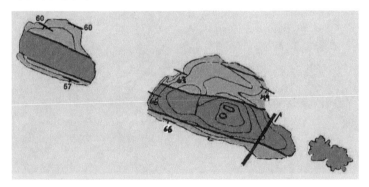

Figure 386. Geologic map of the Cactus Islands.

SPIEDEN ISLAND

Spieden Island (Fig. 387, 388) is an elongate, east–west trending island about three miles long with a relatively straight coastline having no inlets or bays. Water depth drops off sharply offshore.

The island consists entirely of conglomerate, sandstone, shale, and limestone of the Spieden Formation (Fig. 54). Conglomerate makes up about 85 per cent of the rocks on Spieden. Cobbles up to a foot in diameter are common, but most are an inch or less in diameter. Along the northern margin of Spieden Island, about 35 feet of thin-bedded shale contains many fossils. The fossil–bearing beds are overlain by more than 2,000 feet of conglomerate.

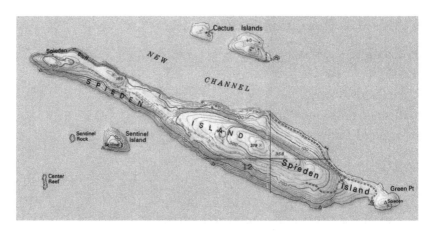

Figure 387. Spieden, Sentinel, and Cactus Islands.

Figure 388. Spieden Island looking east.

Marine sandstone and shale on Spieden Island contain abundant fossils. The oldest fossils occur in the late Jurassic Spieden Bluff Formation, a 65–foot thick bed of sandstone and shale. The Jurassic beds are overlain by fossil-bearing sandstone and shale of the Sentinel Island Formation containing a dozen species typical of lower Cretaceous age. The lower fossil–beds are overlain by 2000 feet of coarse conglomerate. The rocks on Spieden Island contain radioactive isotopes that allow their age to be measured. The Sentinel Island Formation contains mineral grains with an isotope age of 177 million years (early Cretaceous).

Geologic Structure

All of the conglomerate and sandstone beds on Spieden Island have been tilted to the south (Figs. 389, 390). The island is essentially a tilted block of resistant rock. Cactus and Flat Top Islands, north of Spieden Island, rise above sea level from a submerged oval–shaped ridge on the sea floor (Fig. 391). Sandstone and conglomerate on these islands parallel beds on Spieden and dip in the same direction, but are younger, indicating that they are separated by a fault (Fig. 392).

Figure 389. Steeply-dipping sandstone beds tilted to the south
at the western end of Spieden Island.

Figure 390. South-dipping sandstone beds at the eastern end of Spieden Island.

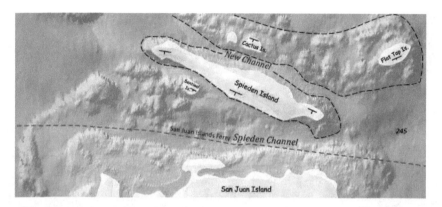

Figure 391. Bathymetry of the sea floor in the vicinity of Spieden Island. Beds of conglomerate and sandstone on Cactus and Flat Top Islands rest on a submerged ridge on the sea floor and dip in the same direction as beds on Spieden Island but are younger so a fault in New Channel must separate them.

Figure 392 is a geologic N–S cross–section from San Juan Island to the Cactus Islands showing offsetting of the beds by faulting.

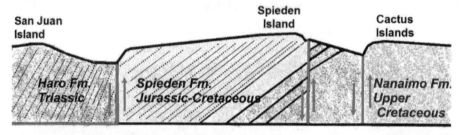

Figure 392. Geologic cross–section from San Juan Island to Cactus Islands.

Figure 393. Faults separating San Juan Island from Spieden Island and Spieden Island from Cactus Islands.

Terraces

A broad terrace extends along the north side of Spieden Island (Figure 394). It stands about 40 feet above sea level at the shoreline and rises to about 100 feet above sea level at the edge of the ridge making up Spieden Island. Most of the terrace appears to be cut on bedrock. Its origin would normally be considered marine, but its broad width cut on bedrock is unusual. A marine terrace occurs at the southeast end of Spieden Island (Fig. 395).

Figure 394. Terrace along the north shore of Spieden Island.
Note the outcrops of sandstone and conglomerate on the right.

Figure 395. Marine terrace on the SE end of Spieden Island.

Glacial erratics, deposited from the mile-thick Cordilleran Ice Sheet that covered the area during the last Ice Age, are scattered about on the ground surface.

Figure 396. Glacial erratics on the south shore of Spieden Island.

SENTINEL ISLAND

Sentinel Island (Fig. 397) is composed of about 800 feet of conglomerate identical in composition to the conglomerate on Spieden Island. At the north end of the island, the beds trend N80°W and dip 45°SW. At the south edge of the island the beds trend N82°W and dip 55°SW.

Figure 397. Sentinel Island sandstone and conglomerate dipping about 55° S

SENTINEL ROCK

Sentinel Rock is a small, bare rock five feet above sea level about 350 yards west of Sentinel Island, connected to Sentinel Island by a submerged reef. It is composed of rocks of the Spieden Formation.

WALDRON ISLAND

Waldron Island (Figs. 398, 399) consists of late Cretaceous Nanaimo sandstone, shale, and conglomerate, mantled by glacial deposits (Fig. 400).

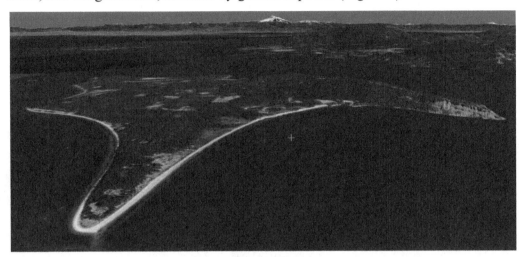

Figure 398. Waldron Island.

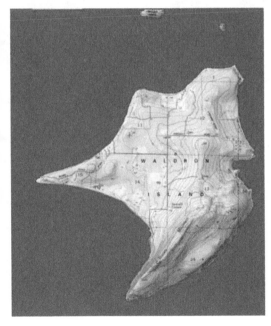

Figure 399. Topographic map of Waldron Island.

Figure 400 Geologic map of Waldron Island.

Nanaimo conglomerate and sandstone make up the SE portion of Waldron Island, including a high ridge with bold cliffs NE of Point Disney (Figs. 399–401). Fossils occur in shaly sandstone. Boulders in the conglomerate are commonly several feet in diameter and consist of granite, andesite, chert, and coarse–grained, dark, igneous intrusive rocks.

Isolated outcrops of sandstone and minor conglomerate are scattered along the north and east coasts of the island (Figs. 402, 403), but the rocks are covered in most places by glacial sediments. Fossil oysters (*Ostrea*) are abundant in the sandstone.

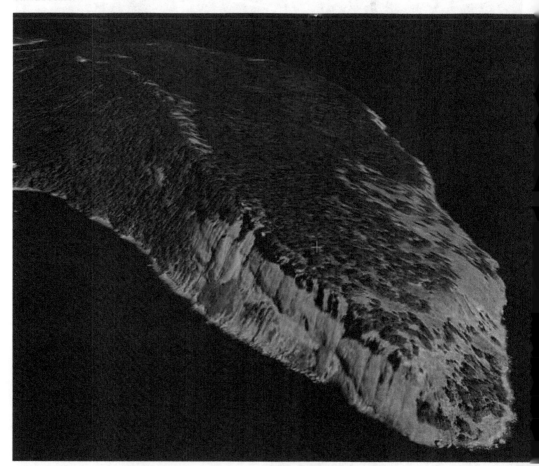

Figure 401. Point Disney, Waldron Island.

A large, thick–shelled species of oyster (*Ostrea*) occurs with other fossils at Point Hammond, near Fishery Point, and Bare Island. The richest fossil–bearing horizon on Waldron Island occurs just above the Point Disney conglomerate.

Figure 402. Rocky points of Nanaimo sandstone at Mail Bay, Waldron Island.

Figure 403. Dipping Nanaimo beds Otter Cove, Waldron Island.

The NE half of Waldron Island is mantled with glacial sediments (Fig. 400) with scattered rocky points of sandstone poking out from underneath. The glacial sediments are exposed in sea cliffs along the shoreline. Near Point Hammond at the northeast corner of the island, sea cliffs of glacial deposits rise to an elevation of 100 feet (Fig. 404).

Figure 404. Ice Age glaciomarine sediments and glacial till overlying outwash sand and gravel, northern Waldron Island.

Sonar bathymetry of the sea floor indicates that the Waldron fault marks the SE margin of the island and extends northward.

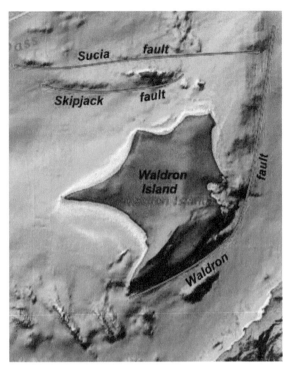

Figure 405. Sea floor faults near Waldron Island. (Base map by NOAA)

SKIPJACK ISLAND

Skipjack Island north of Waldron Island consists of tilted beds of Nanaimo sandstone dipping steeply to the north (Fig. 406) on an uplifted fault block bounded on the north by the Sucia fault and on the south by the Skipjack fault (Fig. 405).

Figure 406. Nanaimo sandstone dipping steeply to the north, Skipjack Island.

BARE ISLAND

Bare Island (Fig. 407) lies just north of Waldron Island. It consists of Nanaimo sandstone, shale, and conglomerate trending N70–80°W and dipping 80° N.

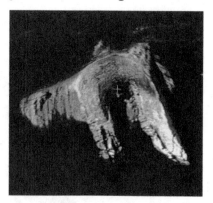

Figure 407. Bare Island. Nanaimo sandstone dipping steeply to the north

PATOS ISLAND

Patos Island (Fig. 408, 409) is composed of Chuckanut sandstone and shale dipping 40–60°N (Fig. 410.) The island is made of two ridges of resistant sandstone (Fig. 408, 411) that bend around slightly at the eastern end of the island (Fig. 412), making a shallow syncline.

Figure 408. Patos Island.

Figure 409 Patos Island.

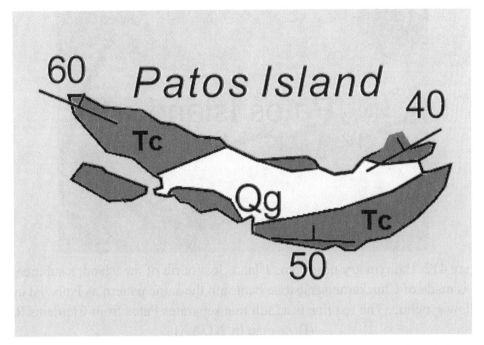

Figure 410. Geologic map of Patos Island. Numbers are angles of dip of the beds. Tc=Chuckanut sandstone, Qg=glacial sediments. (Modified from Lappen 2000)

Figure 411. Resistant Chuckanut sandstone beds protruding from the east end of the island. The bays between ridges consists of weak shale etched out by erosion. The beds dip to the north (right) and are bent into a syncline whose axis is tilted to the north.

Figure 412. Bathymetry near Patos Island. Just north of the island, a submerged reef is made of Chuckanut sandstone bent into the same pattern as Patos Island–a shallow syncline. The red line is a fault that separates Patos from Clements Reef. (Base map by NOAA)

LUMMI ISLAND

The topography of Lummi Island (Figs. 413, 414) is rather unusual. The southern half of the island rises quickly from sea level to 1740 feet, whereas the northern half is low and flat (Fig. 415). The entire southern half of the island is a large ridge of rock tilted to the east parallel to the shoreline. Sheer cliffs on the southwest side of the island rise abruptly from sea level and talus slopes extend from the shoreline to 1000 feet, forming the Devils Rock Slide. The northeastern side of the island slopes more gently eastward parallel to the dip of the rocks.

Figure 413. Lummi Island.

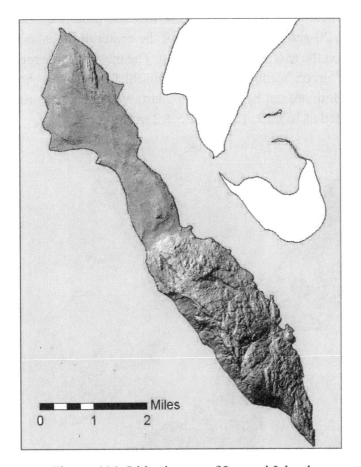

Figure 414. Lidar image of Lummi Island

Figure 415. Northern Lummi Island made of Chuckanut sandstone mantled with
Ice Age glaciomarine sediments and raised beach deposits.

The oldest rocks on Lummi Island are metamorphic and igneous rocks that occur at a small, 20-acre, isolated knob near the center of the island (Fig. 416) and at Lovers Bluff on the east side of Legoe Bay. The metamorphic rocks at the central knob consist of green-hornblende-rich greenschist, which has been intruded by granitic dikes. Both are cut by numerous quartz veins. Granitic rocks at the knob were isotope dated at 163.2 ± 4 and 162.5 ± 2 million years.

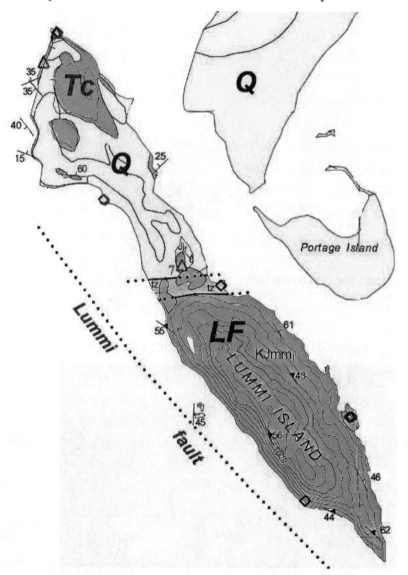

Figure 416. Geologic map of Lummi Island. LF=Lummi Fm., Tc=Chuckanut Fm., Q=Ice Age and modern sediments. (Modified from Lappen, 2000)

Carter Point Formation

Almost all of the high southern portion of Lummi Island is made up of highly indurated greywacke sandstone, argillite, and conglomerate/breccia about 4500 feet thick. Massive beds of conglomerate 150-200 feet thick form high, vertical cliffs along the southwest side of the island. The beds are not laterally continuous for any great distance before grading into a different composition (sandstone, shale, conglomerate). Thus, lateral correlation of stratigraphic sections is difficult.

A stratigraphic section measured by Calkin (1959) on the southwest side of the island (Fig. 41) shows more than 1000 feet of sandstone, shale, and conglomerate. The lower ~800 feet consist mostly of sandstone and shale with very little conglomerate. The upper 300 feet of the section has a lot of conglomerate.

Cross bedding and graded bedding are common in both the sandstone and conglomerate. The greywacke sandstone is extremely well indurated and thus, quite resistant to erosion. Because of this, these rocks have been extensively quarried for rip-rap and crushed for road building. The rocks are cut by numerous quartz veins (Fig. 39)

Radiolarian microfossils in chert at the base of the Carter Point greywacke sandstone about a mile and a half NW of Carter Point indicate an age of late Jurassic to early Cretaceous.

Reil Harbor volcanic rocks

Submarine pillow lava and volcanic breccia overlie the Carter Point greywacke sandstone and conglomerate. Outcrops are limited to five relatively small areas, the largest of which is about 50 acres at Reil Harbor on the southeast side of Lummi Island (Fig. 417). At Sunset Cove, the Reil Harbor volcanic rocks overlie the Carter Point sandstone and conglomerate. Other localities include Legoe Bay, Lummi Point, and Migley Point.

Small lenses of limestone 1–3 feet wide are interbedded with the pillow lava. Highly contorted ribbon chert about 60 feet thick is interbedded with the lavas at Reil Harbor and Migley Point. All of these rocks have been altered and recrystallized. Dense, dark colored, volcanic breccia is interbedded with the pillow lavas at Sunrise Cove, Migley Point, and Legoe Bay.

Glassy and crystal–rich volcanic tuff (ash) between the Reil Harbor volcanic rocks and the overlying Chuckanut sandstone is exposed on the east coast north of Lummi Point (Fig. 417).

Radiolarian microfossils in chert interbedded with pillow basalt at Reil Harbor indicate a Jurassic age and radiolarian microfossils in black argillite at Lovers Bluff indicate a middle to late Jurassic age.

268

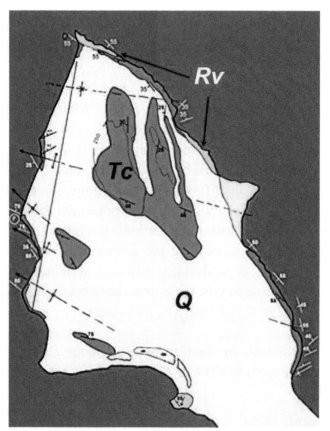

Figure 417. Geologic map of northern Lummi Island showing details of Chuckanut folds and Reil Harbor volcanic rocks . Tc=Chuckanut sandstone, Rv=Reil Harbor volcanic rocks, Q=Ice Age and modern sediments.

Chuckanut sandstone

Chuckanut sandstone makes up almost all of the bedrock on the low northern part of Lummi Island (Fig. 417). It consists of about equal amounts of sandstone and conglomerate which are quite different in composition than the sandstone and conglomerate of the underlying Carter Point Formation. The Carter Point rocks consist of dark, poorly sorted, highly indurated, angular sand and silt composed almost entirely of volcanic particles. In contrast, Chuckanut sandstone consists of light colored, clean, less indurated, rounded sand derived from a granitic source terrain, rather than from a volcanic source.

Chuckanut sandstone rests on Reil Harbor volcanic rocks at Migley Point (Fig. 418). Conglomerate makes up about half of the Chuckanut on Lummi Island. Some conglomerate beds are 30 feet or more thick. The larger particles are mostly rounded pebbles with smaller amounts of cobbles. The pebbles and cobbles consist of about three-fourths quartz, granite, and chert and about one fourth volcanic rocks.

At Fern Point, abundant plant fossils occur in sandstone. Fossil wood still retains identifiable tree rings and an upright fossil log occurs in an upright position of growth.

An isotope age of 53.4 ± 7.1 million years was measured from Chuckanut sandstone just south of Migley Point. This age corresponds well with similar ages from istotope dating of Chuckanut sandstone elsewhere and is consistent with fossil plant ages.

Figure 418. Chuckanut sandstone (tan rock on right) overlying
Jurassic volcanic rocks, Point Migley.

Faulting

Several major faults occur on or near Lummi Island. The Lummi fault, a major fault trending NW–SE along the entire west side of the island, lies just offshore (Fig. 416). Another fault separates the high southern half of the island from the low northern half (Fig. 416). Many smaller faults cut the Carter Point greywacke on the southern part of Lummi (Fig. 419).

Figure. 419. Faults in Carter Point greywacke sandstone on the southern half of Lummi Island.

The Ice Age on Lummi

During the last Ice Age 15-17,000 years ago, the Cordilleran Ice Sheet was at least 6,000 feet thick over Lummi Island. About 15,000 years ago, the cold climate that had resulted in the growth of huge ice sheets suddenly warmed abruptly, causing rapid melting of the ice sheet. By 13,000 years ago, the ice sheet had thinned from 6,000 feet to only a few hundred feet, allowing marine water to enter the depressed lowland and floating the remaining ice. This caused rapid collapse of the Cordilleran Ice Sheet and glaciomarine drift was deposited over the low, northern part of Lummi as a 10–30 feet thick, shell bearing pebbly silt. Good examples of the glaciomarine drift occur in sea cliffs along the coast.

As the Cordilleran Ice Sheet flowed over Lummi Island, it scoured and polished the bedrock and cut grooves parallel to the direction of ice flow (Fig. 420), especially in the high rocky part of the island. The Cordilleran Ice Sheet dropped large numbers of erratic boulders over the island, which are obvious along the shorelines and in open fields. Most of the erratics consist of granite carried to the island from British Columbia.

Figure 420. Glacial grooves on ice–scoured bedrock.

Near the end of the last Ice Age, melting ice floating in marine water deposited glaciomarine pebbly silt over the lower part of Lummi Island (Fig. 416). Fossil marine shells, radiocarbon dated at about 13,000 years, are common in these deposits. After the floating ice had melted away, sea level was several hundred feet higher than present and multiple beach sediments and marine terraces were left over most of the north half of Lummi Island (Fig. 124, 125).

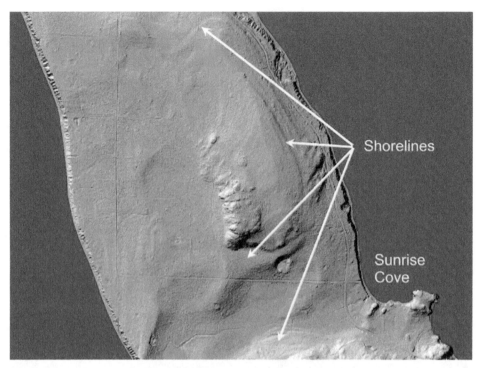

Figure 421. Marine shorelines near Sunrise Cove made during the end of the last Ice Age about 13,000 years ago.

Figure 422. Marine shorelines north of Legoe Bay, Lummi Island.

273

Lummi Rocks

Lummi Rocks (Fig. 423) are composed of Carter Point greywacke sandstone dipping 45°E.

Figure 423. Lummi Rocks off the SW shore of Lummi Island.

CONCLUDING REMARKS

Knowing something about the geology of the San Juan Islands will enhance the reader's enjoyment of the islands with a greater appreciation of their natural beauty. Behind every landform and rock is a geologic story that can be read with a little geologic knowledge.

Many contributions to the geology of the San Juan Islands have been made in the past nine decades since Roy McLellan rowed his boat to all 172 islands and produced the first comprehensive geologic map of the islands. The backbone of the geology of the islands lies in understanding the sequence and chronology of the sedimentary rocks in the area. Virtually all we know about fossil ages of the sedimentary sequences is based on the work of Ted Danner who spent many years collecting and identifying fossils to determine the age of the rocks in which they occur. Joe Vance and John Whetten integrated years of geologic work into a framework that has persisted for 40 years. Darryl Cowan and Mark Brandon built upon this framework, and many others, too numerous to mention here, have also contributed to our understanding of the geology of the islands.

Much of the geologic information in this book comes from new technology in laser, sonar, and satellite technology that has only recently become available. LIDAR reveals many new geologic features that previously could not be observed in any other way. Sonar images of the sea floor allow us to now see geologic structures that were previously unknown and to better understand what can be seen above sea level. Sonar images of the sea floor may be found in the appendix. Satellite images give us a unique perspective of geologic structures and other features that were not available to earlier geologists.

The new data LIDAR, sonar, and satellite data have produced stunning results that require re–evaluation of the fundamental geologic framework of the islands. The new data clearly shows that all five of the postulated thrust faults that comprise the San Juan Thrust System are actually made up of segments of steeply dipping faults and are not actually thrust faults. This doesn't mean that no thrust faulting has occurred in the islands–the older rocks have indeed been intensely sheared under strong compressive crustal forces. What the new data show is that the five postulated, far–traveled, thrust plates do not really exist and the San Juan Thrust System can no longer be considered a valid concept.

Whether you read this book from cover to cover, extract the geologic history from the maps and images, or simply use the book as a reference to look up the geology of specific islands, hopefully you will find in this book the geologic information that you want.

APPENDIX

Bathymetric images of the sea floor (NOAA)
(Green=islands, blue=sea floor)

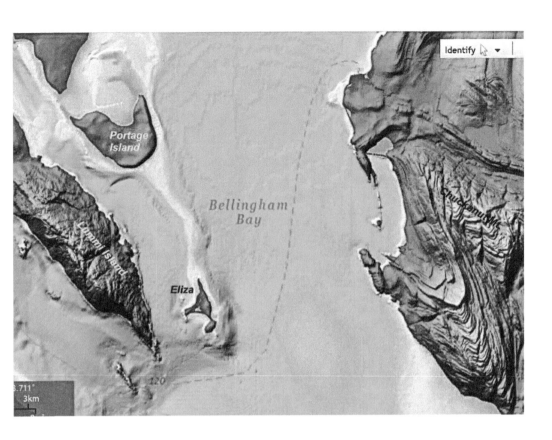

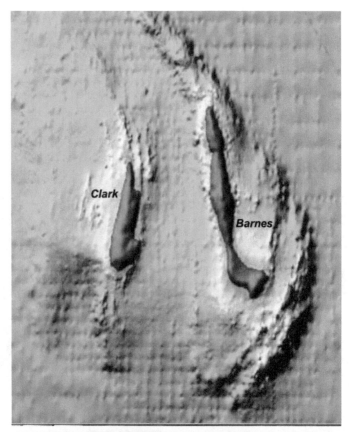

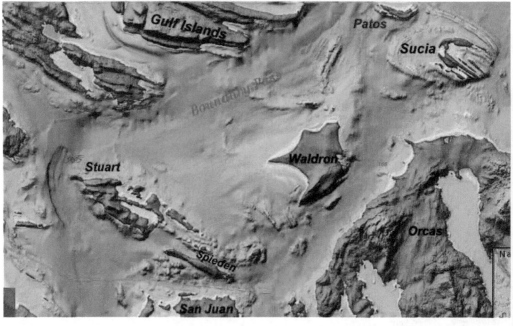

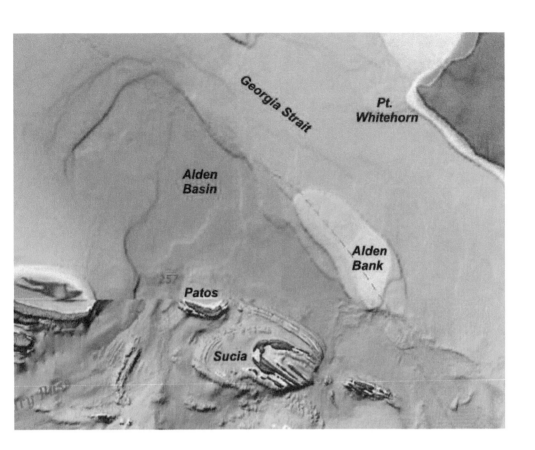

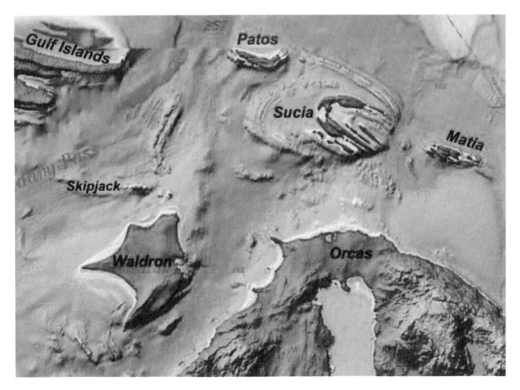

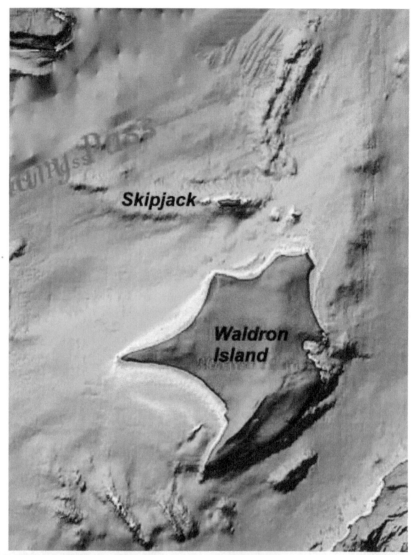

Skipjack

Waldron
Island

Cactus Islands

New Channel

Spieden Island

Flattop

Identify ▾ | Basemap ▾ | Options ▾

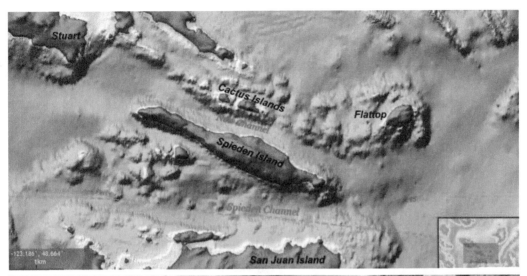

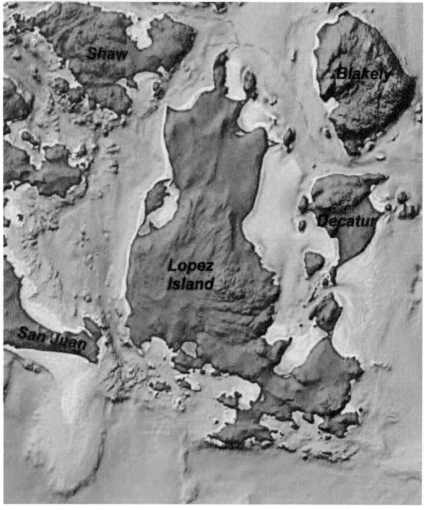

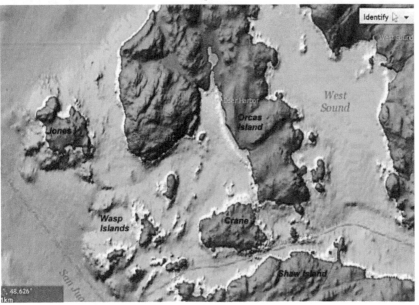

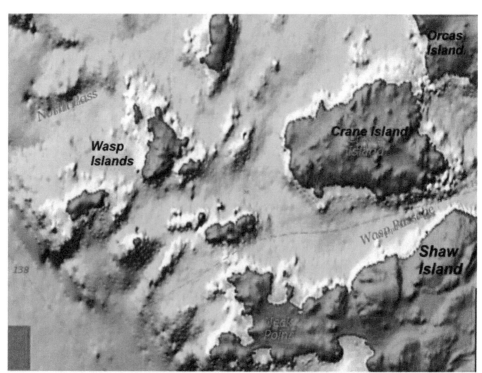

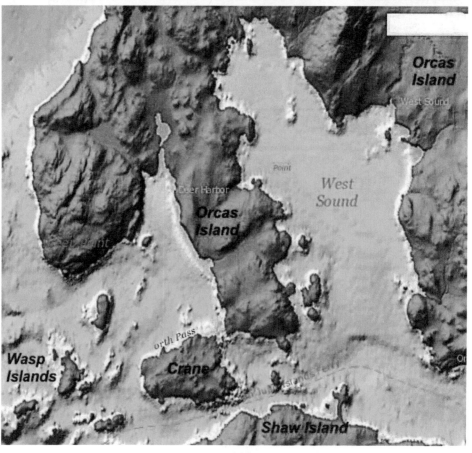

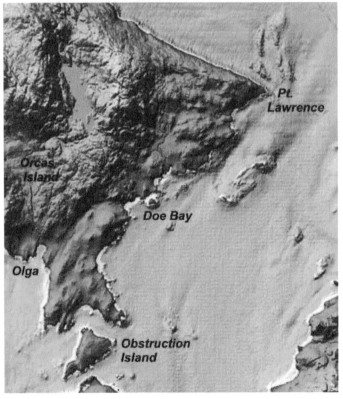

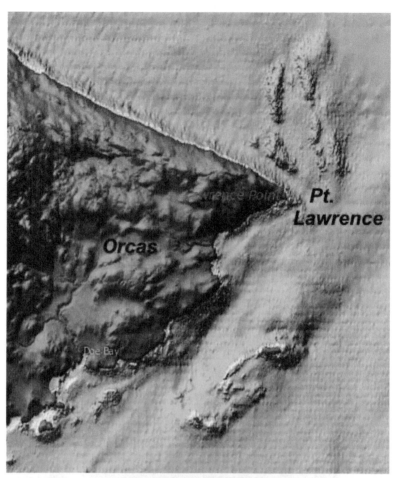

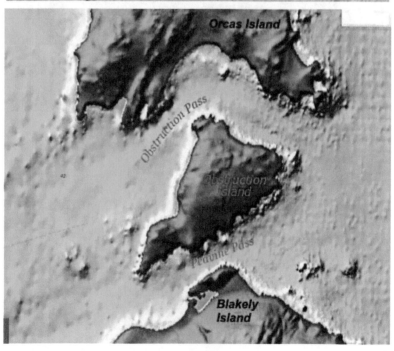

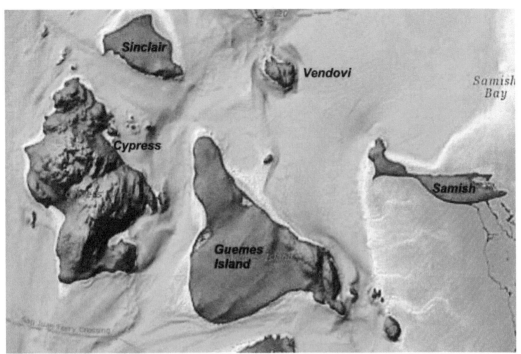

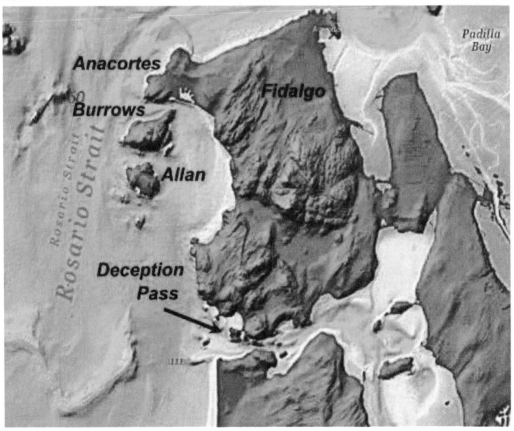

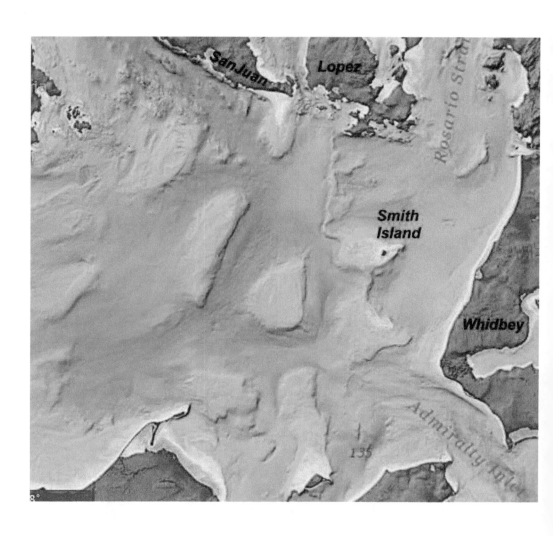

GLOSSARY

Abrasion: The wearing away by friction.

Aggradation: The process of building up a surface by deposition.

Alluvial fan: Low, cone-shaped deposit formed by a stream issuing from mountains into a lowland.

Alluvium. Sand, gravel, and silt deposited by rivers and streams in a valley bottom.

Ammonite. Coiled nautiloid fossils that lived in many of the world's ocean during the Mesozoic Era. They are particularly valuable determining the age of sedimentary rocks in which they occur.

Amphibole. A family of silicate minerals forming prismatic or needlelike crystals that contain iron, magnesium, calcium, and aluminum. Hornblende always has aluminum and is a most common dark green-to-black variety of amphibole, forming in many igneous and metamorphic rocks. Actinolite has no aluminum and is light green. Blue amphibole contains sodium and is bluish in color.

Amphibolite. A metamorphic rock, made mostly of hornblende and plagioclase.

Andesite. Fine-grained, generally dark-colored, volcanic rock. Commonly has visible crystals of plagioclase feldspar. Occurs in lava flows and dikes. The most common rock in volcanic arcs.

Anticline: Structure in which beds dip in opposite directions from the axis.

Aquifer: a formation that is water-bearing.

Argillite. Fine-grained sedimentary rocks made mostly of silt and clay. Includes shale, mudstone, siltstone, and claystone. Commonly black.

Arkose. Sandstone composed of quartz and feldspar formed by erosion of granite and deposition of sand.

Axis of a fold: The line following the apex of an anticline or the lowest part of a syncline.

Axial plane of a fold. A plane through the central part of a fold.

Bar: An embankment of sand or gravel deposited on the floor of a stream, sea, or lake.

Basal till: Poorly sorted mixture of sand, silt, clay, pebbles, cobbles, and boulders deposited from, the base of a glacier.

Basalt. Fine-grained, generally black, volcanic rock relatively rich in iron, magnesium, and calcium. Occurs in lava flows and dikes.

Base level: The level below which a land surface cannot be reduced by running water. Sea level is the ultimate base level for streams.

Batholith. Very large mass of slowly cooled, intrusive molten rock such as granite at least 50 square miles in area.

Bedding. Sedimentary layers in a rock. Beds are distinguished from each other by grain size and composition.

Blueschist. Metamorphic rock rich in blue amphibole formed by high pressure and low metamorphic temperature. Often associated with greenschist.

Breccia. Rock made up of angular fragments. **Volcanic breccia** is made of volcanic rock fragments blown out of a volcano or eroded from it. **Sedimentary breccia** is formed by deposition of angular fragments from water. **Fault breccia** is made by breaking and grinding rocks along a fault.

Carbon-14: A radioactive isotope of carbon with atomic weight 14, produced by collisions between neutrons and atmospheric nitrogen. Used as a natural geologic clock to determine the age of organic material such as wood or shells.

Calcite. Mineral made of calcium carbonate ($CaCOs$). Generally white, easily scratched with knife. Most seashells are calcite or related minerals. Primary component of limestone.

Chemical weathering. Chemical decomposition of rocks by weathering.

Chert. Sedimentary rock made of extremely fine-grained quartz. Commonly made up of millions of globular siliceous skeletons of tiny marine plankton called radiolarians.

Chlorite. Green, platy, metamorphic minerals formed by alteration of amphibole or pyroxene. Common in metamorphic rocks such as greenschist formed by low heat and high pressure.

Cinder cone: A volcanic cone formed by the accumulation of volcanic cinders or ash around a vent.

Clay. Particles less than 1/16 millimeter in diameter. Also a family of platy silicate minerals generally too small to be seen even with a microscope. A common product of rock weathering, especially of rocks containing much feldspar. The term *clay is* also used to refer to very, very fine sedimentary grains whether or not they are made of clay minerals.

Climate: The sum total of the meteorological elements that characterize the average and extreme condition of the atmosphere over a long period of time at any one place or region of the earth's surface.

Conglomerate. Sedimentary rock made of rounded pebbles, cobbles, and boulders greater than at least 2 mm (about 1/13th of an inch) in diameter.

Conodont. Tiny fossil bone material of marine organisms used for age determination of Paleozoic sedimentary rocks.

Continental glacier: An ice sheet covering a large part of a continent.

Contour interval: The difference in elevation between two adjacent contour lines.

Contour: An imaginary line on the surface of the ground, every point of which is at the same altitude.

Coral reef: A reef formed by the action of reef-building coral polyps, which build internal skeletons of calcium carbonate.

Cordilleran Ice Sheet. Ice cap that grew in western North America during the Pleistocene Epoch, beginning in Canada and covering much of British Columbia, Alaska, and northernmost United States.

Creep: The slow downslope movement of rock fragments and soil.

Cross bedding. Beds of sand or gravel deposited at an angle to the horizontal on the backside of sand bars or dunes.

Debris flow: A moving mass of water–lubricated debris.

Deglaciation: The uncovering of an area from beneath glacier ice as a result of shrinkage of a glacier.

Delta: An alluvial deposit, often triangular-shaped, formed where a stream enters the ocean or a lake and drops its load of sand, silt, or gravel.

Differential erosion: The more rapid erosion of portions of the earth's surface as a result of differences in the erodibility of the rock or in the intensity of surface processes.

Differential weathering: When rocks are not uniform in character but are softer or more soluble in some places than in others, an uneven surface may be developed.

Dike. Tabular body of igneous rock formed where molten rock is injected into cracks in pre–existing rock.

Diorite. Intrusive igneous rock made of plagioclase feldspar and amphibole and/or pyroxene. Similar to granite except it has little or no quartz.

Dip: The angle at which a bed or other planar feature is inclined from the horizontal.

Drainage basin: The area drained by a river system.

Emergence: A part of the ocean floor that has become dry land, but does not imply whether the sea receded or the land rose.

End moraine: A ridge-like accumulation of glacial sediment deposited at the terminus of a glacier.

Epidote. Family of silicate minerals common in metamorphic rocks. Apple green to straw yellow.

Erratic: A rock transported by a glacier or by floating ice, different from the bedrock on which it lies

Escarpment: A cliff or relatively steep slope separating gently sloping tracts.

Estuary: A bay at the mouth of a river, where the tide influences the river current.

Facet: A flat surface produced by abrasion on a rock.

Fan: A low cone-shaped accumulation of debris deposited by a stream descending from a ravine onto a plain, where the material spreads out in the shape of a fan.

Fault. A fracture along which there has been displacement of the two sides relative to one another. Abrupt movements on faults cause earthquakes. Where the crack is roughly vertical the rocks may move up or down or sideways or in some combination. If the fault is inclined at a low angle to the Earth's surface and rock on one side of the fault moves up and over rock on the other side, it is a **thrust fault.**

Fault scarp: The bluff formed by offset of the land surface by a fault.

Feldspar. Family of silicate minerals whose crystals are stubby prisms, generally white (plagioclase) or pink (potassium feldspar), containing aluminum, potassium, sodium, and calcium.

Fission tracks. Microscopic tunnels in crystals and glass made by nuclear particles emitted by radioactive elements, usually uranium. The number of fission tracks in glass and zircon crystals increases with time and can be used as a dating method..

Floodplain: A strip of relatively smooth land on a valley floor bordering a stream, built of sediment deposited during times of flooding.

Fluting: Smooth deep furrows worn in the surface of rocks by glacial or stream erosion.

Fluvial: Pertaining to rivers or produced by river action.

Foliation. Parallel arrangement of minerals, especially platy minerals such as micas, in a rock, so as to give it a foliated look, like pages in a book. Foliated rocks tend to break along the foliation and form slabs. Mostly found in metamorphic rocks.

Foraminifera. Microscopic marine organisms that make tiny shells of calcite. Useful as guide fossils for age determination of sedimentary rocks.

Fossil. Once–living organism whose hard parts are preserved when buried in sediment.

Fusulinid foraminifera. Microscopic marine fossils useful for determination of the age of late Paleozoic sedimentary rocks.

Gabbro. A dark coarse-grained rock igneous rock chemically equivalent to basalt but crystallized slowly at great depth below the Earth's surface.

Geomorphology: The study of physical and chemical processes that affect the origin and evolution of surface forms.

Glacial drift: All sediment deposited directly or indirectly from a glacier or by its meltwater.

Glacial groove: Elongate groove carved in rock by an overriding glacier.

Glacial striae: Scratches on smoothed surfaces of rocks made by glacial erosion.

Glacial till: Poorly sorted mixture of sand, silt, clay, pebbles, cobbles, and boulders deposited from a glacier.

Glacier: A body of ice, firn, and snow, originating on land and showing evidence of past or present flow.

Glaciomarine deposits. Poorly sorted sediments deposited from ice floating in the sea.

Gneiss. A light-colored, coarse grained metamorphic rock made by recrystallization of older rocks with chemical compositions similar to granite.

Graben: A fault block down-dropped relative to the rocks on either side.

Graded bedding: Progressive decrease in grain size upward in sediment as a result of deposition from turbidity currents in which the coarsest grains settle out first.

Gradient: Slope expressed as the angle of inclination from the horizontal.

Granite. A coarse-grained igneous rock made of feldspar and quartz that has crystallized from molten rock at great depth below the Earth's surface where crystallization is slow and the minerals are large.

Granitic rocks. A general term for coarse grained igneous rocks composed mostly of feldspar and quartz.

Greenschist. Green, foliated, low grade metamorphic formed by recrystallization of basalt under low temperature and high pressure. Composed of the green minerals chlorite, actinolite, and epidote, which make the rock green.

Greenstone. A green metamorphic rock made by recrystallization of basalt or chemically equivalent rocks. Composed of the green minerals chlorite, actinolite, and epidote, which make the rock green.

Greywacke. Sandstone composed of angular volcanic grains. Usually poorly sorted.

Groundwater: Subsurface water.

Headland: A projection of the land into the sea, as a peninsula or promontory.

High–angle fault: A fault in which the angle of dip is steep.

Hornblende. A mineral of the amphibole group. Usually black or very dark green.

Hornblende schist. A schist rich in hornblende. Generally with abundant plagioclase feldspar. Grades into amphibolite.

Ice Age: Time when huge ice sheets covered most high latitude areas.

Ice sheet: A large glacier of continental proportions forming a continuous cover over a land surface.

Igneous intrusions: Injection of molten rock into pre–existing rocks.

Igneous rocks. Rocks formed by crystallization of molten rock.

Interglacial: Pertaining to nonglacial time between glaciations.

Intrusion. Injection of molten rock into other pre–existing rocks.

Island arc. A chain of oceanic volcanic islands at a subduction zone. The Aluetian Islands are a good example.

Isostatic uplift: Uplift of the land surface as a result of unloading of the weight of massive ice sheets.

Isotope. Elements with slightly different numbers of neutrons in their nucleus than is usual for a particular element. For example, carbon–14 is an isotope of carbon because most carbon has 12 neutrons in its nucleus, but carbon–14 has 14 neutrons in its nucleus and is radioactive.

Isotope dating. Determination of the age of a rock by measuring specific isotopes.

Joint: A fracture in a rock.

Kettle: A depression in glacial drift, made by the melting of a detached mass of glacier ice that has been either wholly or partly buried in the drift.

Lacustrine: Pertaining to lakes.

Landslide: Downslope sliding of earth.

Lava. Molten rock that has flowed out onto the Earth's surface.

LIDAR: Image of the Earth's surface produced by laser beams.

Limestone: A sedimentary rock composed of the mineral calcite (calcium carbonate--$CaCO_3$) commonly formed from calcium carbonate shells of marine creatures or by crystallization of calcite from sea water.

Lineation. Parallel arrangement of elongate minerals or groups of minerals, like pencils parallel to one another.

Lithosphere. Outer shell of the Earth made of the crust and the uppermost part of the mantle.

Longshore drifting: The movement of sediment parallel to the shore by waves approaching a shoreline obliquely.

Magma. Molten rock formed deep in the Earth. When magma pours out on the Earth's surface it is called **lava.**

Mantle. Interior part of the Earth surrounding the core and below the crust. Made up of dense, iron- and magnesium-rich (ultramafic) rock such as dunite and peridotite.

Marble. Metamorphic rock of calcium carbonate made by recrystallization of limestone by heat.

Marine-built terrace: A bench built by deposition of sediment seaward from the shoreline.

Marine-cut terrace: Flat eroded surface made by marine erosion.

Mass wasting: The downslope movement of rock debris under the influence of gravity.

Melange. Mixture of rocks formed by tectonic disruption, such as multiple faulting, which brings disparate rock types together.

Metamorphic rocks. Rocks formed by recrystallization of older rocks by heat and pressure.

Mica. Group of flat, plate-like silicate minerals, which cleave into smooth, flat flakes. **Biotite** is dark black. **Muscovite** is light-colored.

Mineral. A naturally occurring, crystalline compound having definite chemical and physical properties

Moraine. A ridge of glacial sediment along the margin of a glacier composed of rock debris. An **end moraine** forms at the terminus of a glacier. A **terminal moraine** is an end moraine at the farthest advance of the glacier. A **lateral moraine** forms along the sides of a glacier.

Mudflow: A viscous flowage of mud and sediment lubricated with water.

Nappe. Thrust faults having large displacement, often many miles.

Oceanic rocks. Rocks formed in the ocean.

Olivine. A green, glassy mineral formed at high temperature. Common in basalt, especially ocean-floor basalt and in ultrarmafic rocks.

Ophiolite. Mixture of rocks often associated with island arcs.

Outcrop. Exposure of rocks at the surface.

Outwash. Glacial deposit of sand, silt, and gravel formed downstream from a glacier by meltwater streams and rivers.

Outwash plain: A topographic plain made by deposition of sand and gravel by meltwater streams from a glacier.

Oxbow: A crescent-shaped lake formed in an abandoned river bend by a meander cutoff.

Paleomagnetism. The magnetism of a rock imparted to it by the Earth's magnetic field when the rock formed.

Paleontology. The study of fossils.

Pegmatite. Generally very coarse-grained rocks of feldspar and quartz. Crystals are usually over an inch acrossand in some pegmatites, feldspar crystals are several feet across.

Peridotite: Rock made of olivine and pyroxene.

Petrology. Study of rocks.

Phyllite. A very fine-grained, foliated metamorphic rock, generally derived from shale or fine-grained sandstone. Phyllites are usually black or dark gray; the foliation is commonly crinkled or wavy. Differs from less recrystallized slate by its sheen, which is produced by barely visible flakes of muscovite (mica).

Pillow basalt. Lava containing many bulbous, ellipsoidal shapes as a result of eruption under water.

Plankton. Tiny, microscopic animals or plants that live floating in water.

Pleistocene. The last Ice Age.

Pluton. Body of igneous of rock that crystallized from molten rock deep in the earth.

Pre–Cambrian: Earliest period of the Earth's history from about 4 ½ billion to 540 million years ago.

Prehnite. Green metamorphic mineral formed under low temperature and pressure.

Pyrite. Yellow, brassy metallic cubes of iron sulfide mineral (FeS). Known as fool's gold.

Pyroclastic. Volcanic rock formed by explosive eruptions and flowage of fragmental material, magma, and gas downslope

Pyroxene. Family of dark green silicate minerals common in basalt and gabbro.

Quartz. Glassy–looking silicon dioxide (SiO_2). One of the most common minerals in the Earth's crust found in granite, veins, and sandstone.

Quartzite. Metamorphosed quartz sandstone or chert.

Quaternary: Geologic time period of Ice Ages during the past two million years.

Radiocarbon age. The age of organic material determined by the amount of the radioactive carbon isotope carbon-14 in wood or shells.

Radiolarians: Single-celled planktonic animals with skeletons of silica.

Radiolarian chert. A rock made up of the siliceous shells of radiolarians.

Recessional moraine: End moraine formed by a stillstand of ice during recession of a glacier.

Reef: A ridge of coral and shell debris formed in warm, shallow seawater.

Reverse fault: A fault in which the hanging wall has moved up relative to the footwall.

Rhyolite. A volcanic rock chemically equivalent to granite but erupted on the land surface. Usually light-colored and fine-grained with tiny, visible crystals of quartz in a dense matrix.

Ribbon chert. Alternating beds of chert and thin shale resembling parallel ribbons.

Sandstone: Rock made of sand grains converted to rock by cementation.

Scarp: A cliff or steep slope.

Schist: Foliated metamorphic rock usually derived by recrystallization from shale by heat and pressure.

Sea cliff: Bluff along a shoreline made by wave erosion.

Sedimentary rocks. Rocks formed by (1) deposition and cementation of rock particles, (2) by precipitation of chemicals in oceans or lakes, or (3) by accumulation of shells or other organic material.

Serpentine. Low-temperature metamorphism of minerals in ultramafic rocks to form green, greasy-looking, silicate minerals that are slippery to the touch.

Shale. Sedimentary rock form by deposition of mud on the floor of oceans or lakes..

Shearing. Planar stress within a rock that produces sliding planes similar to a deck of cards.

Silicate. The chemical unit silicon tetroxide, SiO_4, that is the fundamental building block of **silicate minerals.** Silicate minerals make up most rocks at the Earth's surface.

Sill. A planar igneous intrusion parallel to pre–existing rock structures such as bedding.

Slump: The downward slipping of a mass of rock or unconsolidated material, usually with backward rotation.

Soapstone. The mineral talc, a very soft, platy mineral. Can be easily carved with a knife.

Spit: A sandbar projecting into a body of water from the shore.

Stagnant ice: A glacier in which the ice has ceased to move.

Striations: Scratches or small grooves.

Strike. The compass direction of a horizontal line on a bedding plane.

Subduction. Process of one crustal plate riding over another crustal plate as the two converge. The **subduction zone** is the area between the two plates, somewhat like a giant thrust fault.

Subglacial: Beneath a glacier.

Submarine fan. Fan- or cone-shaped accumulation of sedimentary debris-sand, gravel, mud-under the ocean along the edge of the land, either a continent or a volcanic arc. Fans may be a few miles to a hundred or so miles across.

Submergence: Inundation by the sea without implication as to whether the sea level rose or the land subsided.

Syncline: A fold in which the beds dip inward from both sides toward the axis.

Talc: A very soft, magnesium silicate mineral, commonly called soapstone because of its softness and slippery feel.

Talus: An accumulation of loose rock at the base of a cliff.

Terminal moraine: A ridge of glacial deposits marking the farthest advance of a glacier.

Terrace. Flat, gently inclined, or horizontal surface bordered by an escarpment. Can be either depositional or erosional.

Terrane. A rock formation or assemblage of rock formations that share a common geologic history. A geologic terrane is distinguished from neighboring terranes by its different history, either in its formation or in its subsequent deformation and/or metamorphism.

Thrust fault: A low-angle, reverse fault that pushes older rocks over younger rocks.

Thrust sheet: Slab of rock pushed over other rocks by tectonic forces.

Till. Poorly sorted, nonstratified rock debris deposited by a glacier.

Tombolo: A former island that has been tied to the mainland by seaward building of a spit.

Tuff. Volcanic ash.

Turbidites. Sediments deposited from muddy slurries on the sea floor. Commonly occurring on the sea floor near island arcs.

Ultramafic rock. Rock very rich in pyroxene and olivine and higher iron and magnesium with lower silicon and aluminum than most crustal rocks. Igneous varieties are peridotite and dunite. May come from the Earth's mantle. A common metamorphic variety is serpentinite.

Unconformity. A gap in a succession of sedimentary beds caused by erosion or non-deposition.

Vein: Tabular rock filling of fractures in rock with minerals precipitated from hot solutions.

Volcanic arc. Arcuate chain of volcanoes formed above a subducting plate.

Volcanic ash: Volcanic rock fragments, glass, pumice, and mineral crystals

Volcanic rocks: Rocks formed at the Earth's surface by volcanic eruptions.

Water table: The upper surface of the saturated zone of ground water.

Wave-built bench: Gently sloping bench built by wave and current deposition of sediment.

Wave-cut bench: Beveled bedrock surface produced by wave erosion.

Weathering: Disintegration and decomposition of rocks by surface processes.

Zircon: Mineral composed of zirconium, silicon, and oxygen in igneous rocks.

REFERENCES

Armstrong, J.E., Crandell, D.R., Easterbrook, D.J., and Noble, J. B. 1965, Late Pleistocene stratigraphy and chronology in southwestern British Columbia and northwestern Washington: Geological Society of America Bulletin, vol. 76, p.321-330.

Atkin, S.A., 1972, Submarine volcanic rocks on the west coast of San Juan Island, Washington: Master of Science thesis, University of Washington, 21 p.

Baichtal, J.F., 1982, The geology of Waldron, Bare, and Skipjack islands, San Juan County, Washington: Master of Science thesis, Washington State University, 222 p.

Belanger, T.I., 2008, Structural geology of the central San Juan Islands, northwest Washington: Master of Science thesis, Western Wash. University, 145 p.

Bergh, S.G., 2002, Linked thrust and strike-slip faulting during Late Cretaceous terrane accretion in the San Juan thrust system, Northwest Cascade orogen, Washington: Geological Society of America Bulletin, vol. 114, p. 934–949.

Blake, C. and Engebretson, D., 2007, Murrelets and molasse in the eastern San Juan Islands: in Floods, faults, and fire; geological field trips in Washington State and southwest British Columbia, Stelling, P. and Tucker, D.S., eds., Geological Society of America Field Guide, vol. 9, p. 137-142.

Blunt, D., Easterbrook, D.J., and Rutter, N.A., 1987, Chronology of Pleistocene sediments in the Puget Lowland, Washington: Washington Division of Geology and Earth Resources Bulletin 77, p. 321-353.

Brandon, M. T. Cowan, D. S., Muller, J. E., Vance, J. A., 1983, Pre-Tertiary geology of San Juan Islands, Washington and Southeast Vancouver Island, British Columbia:, Geological Association of Canada, Mineralogical Association of Canada and Canadian Geophysical Union Field trip guidebook, vol. 1, 65 p.

Brandon, M.T., 1980, Structural geology of Middle Cretaceous thrust faulting on southern San Juan Island, Washington: Master of Science thesis, University of Washington, 130 p.

Brandon, M.T., 1989, Geology of the San Juan–Casades nappes, northwestern Cascade Range and San Juan Islands: Geologic Guidebook for Washington and adjacent areas, Washington Division of Geology and Earth Resources, Information Circular 86, p. 137–162.

Brandon, M.T., Cowan, D.S., and Vance, J.A., 1988, The Late Cretaceous San Juan thrust system, San Juan Islands, Washington: Geological Society of America Special Paper 221, 81 p.

Brandon, M.T., Cowan, D.S., Muller, J.E., Vance, J.A., 1983, Pre-Tertiary geology of San Juan Islands, Washington and Southeast Vancouver Island, British Columbia: Geological Association of Canada, Mineralogical Association of Canada and Canadian Geophysical Union Field trip guidebook, vol. 1, 65 p.

Brandon, M.T., Cowan, D.S., Vance, J., 1988, The Late Cretaceous San Juan thrust system, San Juan Islands, Washington: Geological Society of America Special Paper, vol. 221, 81 p.

Brown, E.H. and Gehrels, G.E., 2007, Detrital zircon constraints on terrane ages and affinities and timing of orogenic events in the San Juan Islands and North Cascades, Washington: Canadian Journal of Earth Sciences, vol. 44, p 1375–1396.

Brown, E.H., 2012, Obducted nappe sequence in the San Juan Islands–northwest Cascades thrust system, Washington and British Columbia: Canadian Journal of Earth Sciences, vol. 49, p. 796–819.

Calkin, P.E., 1959, The geology of the Lummi and Eliza islands, Whatcom County, Washington: Master of Science thesis, Univ. of British Columbia, 140 p.

Carsten, F.P., 1982, Petrology of the Upper Cretaceous strata of Orcas Island, San Juan County, Washington: MS thesis, Washington State University, 120 p.

Cowan, D.S., Brown, E.H.,Whetten, J.T., 1977, Geology of the southern San Juan Islands: in Brown, E.H. and Ellis, R.C., eds, Geological Society of America, Geological excursions in the Pacific Northwest, p. 309-338.

Dainty, N.D., 1981, Petrology of the strata of Patos Island, San Juan County, Washington: Master of Science thesis, Washington State University,107 p.

Danner, W.R., 1957, A stratigraphic reconnaissance in northwestern Cascades and San Juan Islands of Washington State: PhD thesis, University of Washington, 562 p.

Danner, W.R., 1966, Limestone resources of western Washington, with a section on the Lime Mountain deposit by G.W. Thorsen: Washington Division of Mines and Geology Bulletin vol. 52, 474 p.

Danner, W.R., 1977, Paleozoic rocks of northwest Washington and adjacent parts of British Columbia: in Paleozoic paleogeography of the western United States. J.H. Stewart, C.H. Stevens, and A.E. Fritsche eds., Society of Economic Paleontologists and Mineralogists Pacific Section, Pacific Coast Paleogeography Symposium vol. 1, p. 481–502.

Dean, P.A., 2002, Paleogeography of the Spieden Group, San Juan Islands, Washington: Master of Science thesis, Western Washington University, 59 p.

Dethier, D.P., White, D.P., Brookfield, C.M., 1996, Maps of the surficial geology and depth to bedrock of False Bay, Friday Harbor, Richardson, and Shaw Island 7.5-minute quadrangles, San Juan County, Washington: Division of Geology and Earth Resources, Department of Natural Resources Open-File Report, 7 p.

Easterbrook, D. J., 1994, Chronology of Pre-late Wisconsin Pleistocene the Puget Lowland, Washington: in Lasmanis, R., and Cheney, E. S., Regional Geology of Washington State, Washington Division of Geology and Earth Resources, Bulletin 80, p. 191-206.

Easterbrook, D.J. and Rahm, D.A., 1970, Landforms of Washington: Union Printing Co., 156 p.

Easterbrook, D.J., 1958, Geology of eastern Orcas Island and Sucia Island, Washington: University of Washington, Dept. of Geology Report, 70 p. with geologic map.

Easterbrook, D.J., 1963, Late Pleistocene glacial events and relative sea-level changes in the northern Puget Lowland, Washington Geological Society of America Bulletin, vol. 74, p.1465-1483.

Easterbrook, D.J., 1966, Radiocarbon chronology of late pleistocene deposits in northwest Washington: Science, vol. 152, p.764-767.

Easterbrook, D.J., 1968, Pleistocene stratigraphy of Island County: Water Supply Bulletin, Washington, Department of Water Resources, Part 1, vol. 25.

Easterbrook, D.J., 1969, Pleistocene chronology of the Puget lowland and San Juan Islands, Washington: Geological Society of America Bulletin, vol. 80, p.2273–2286.

Easterbrook, D.J., 1976, Quaternary geology of the Pacific Northwest: Dowden, Hutchinson & Ross Stroudsburg, PA, p.441-462.

Easterbrook, D.J., 1979, The last glaciation of Northwest Washington: Pacific Coast Paleogeography Symposium, Society of Economic Paleontologists and Mineralogists, p.177-189.

Easterbrook, D.J., 1992, Advance and retreat of Cordilleran ice sheets in Washington, U.S.A.: Geographie Physique et Quaternaire, vol. 46, p. 51-68.

Easterbrook, D.J., 1999, Surface processes and landforms: Prentice-Hall, 546 p

Easterbrook, D.J., 2003, Cordilleran Ice Sheet glaciation of the Puget Lowland and Columbia Plateau and alpine glaciation of the North Cascade Range, Washington: Geological Society of America Field Guide 4, p. 137–157.

Easterbrook, D.J., 2003, Cordilleran Ice Sheet glaciation of the Puget Lowland and Columbia Plateau and alpine glaciation of the North Cascade Range, Washington: in Easterbrook, D.J., ed., Quaternary Geology of the United States, International Quatenary Association, 2003 Field Guide Volume, Desert Research Institute, Reno, NV, p. 265-286

Easterbrook, D.J., 2010, A walk through geologic time from Mt. Baker to Bellingham Bay, WA: Chuckanut Editions, Bellingham, WA, 329 p.

Easterbrook, D.J., 2014, Multiple Younger Dryas and Allerød moraines (Sumas Stade) and late Pleistocene Everson glaciomarine drift in the Fraser Lowland: in Dashtgard, S., and Ward, B., eds., Trials and Tribulations of Life on an Active Subduction Zone: Field Trips in and around Vancouver, Canada: Geological Society of America Field Guide 38, p. 79–99.

Easterbrook, D.J., 2014, The San Juan Thrust (Nappe) System: new perspectives from LIDAR, sonar, and satellite imagery: Geological Society of America Abstracts.

Easterbrook, D.J., 2015, The San Juan Thrust (Nappe) System: new perspectives from LIDAR, sonar, and satellite imagery: Earth Science and Engineering, 28 p.

Easterbrook, D.J., Crandell, D.R., and Leopold, E.B. 1967, Pre-Olympia Pleistocene stratigraphy and chronology in the central Puget Lowland, Washington: Geological Society of America Bulletin, vol. 78, p.13-20.

Easterbrook, D.J., in press, Late Quaternary Glaciation of the Puget Lowland, North Cascade Range, and Columbia Plateau, Washington: University of Washington Press, Seattle, WA.

Eddy, P., 1975, Quaternary geology and ground-water resources of San Juan County, Washington: in Russell, R.H., editor, Geology and water resources of the San Juan Islands, Washington, Washington State Department of Ecology, Water-Supply Bulletin 46, p. 21-39.

Egem R.J., 1981, Upper Cretaceous stratigraphy of Matia, Clark, and Barnes islands, San Juan County, Washington: Master of Science thesis, Washington State University, 149 p.

Egemeier, R.J., 1981, Upper Cretaceous stratigraphy of Matia, Clark and Barnes Islands, San Juan County, Washington: Master of Science Thesis, Washington State University, 149 p.

Gillaspy, J.R., 2005, Brittle deformation in an ancient accretionary prism setting; Lopez structural complex, San Juan Islands, NW Washington: Master of Science thesis, Western Washington University, 133 p

Hansen, Henry 1943, A pollen study of two bogs on Orcas Island, of the San Juan Islands, Washington. Paul. Bulletin of the Torrey Botanical Club, May, , Vol. 70, Issue 3, p. 236-243. New York Botanical Garden : Bronx, NY, p. 236-243.:

Janbaz, J.E., 1972, Petrology of the Upper Cretaceous strata of Sucia Island, San Juan County, Washington: Master of Science Thesis, Washington State University, 104 p.

Johnson, S.Y., 1978, Sedimentology, petrology, and structure of Mesozoic strata in the northwestern San Juan Islands, Washington: . Master of Science thesis, University of Washington , 105 p.

Johnson, S.Y., 1981, The Speiden Group: An anomalous piece of the Cordilleran paleogeographic puzzle: Canadian Journal of Earth Sciences, vol. 14, p. 2565–2577.

Johnson, S.Y., 1984, Stratigraphy, age, and paleogeography of the Eocene Chuckanut Formation, northwest Washington, Canadian Journal of Earth Sciences, vol. 21, p. 92–106.

Johnson, S.Y., Zimmermann, R.A., Naeser, C.,W., Whetten, J.T., 1986, Fission-track dating of the tectonic development of the San Juan Islands, Washington: Canadian Journal of Earth Sciences, vol. 23, p. 1318-1330.

Lamb, R.M., 2000, Structural and tectonic history of the eastern San Juan Islands, Washington: Master of Science thesis, Western Wash. University. 236 p.

McClellan, R.D., 1927, Geology of the San Juan Islands, Washington: Ph.D. thesis, University of Washington, 185 p.

Meek, F.B., 1856, On Cretaceous fossils from Vancouver and Sucia Islands: Albany Institute, Transactions, v. 4, p. 37-39.

Meek, 1876, Descriptions and illustrations of fossils from Vancouver and Sucia islands, and other northwestern localities: U. S. Geological Survey, Territories Bulletin, v. 2, p. 351-374.

Mercier, J.M., 1977, Petrology of the Upper Cretaceous strata of Stuart Island, San Juan County, Washington: Master of Science Thesis, Washington State University, 157

Muller, J.E., and Jeletzky, J.A.,, 1978, Revisions to the stratigraphy and biochronology of the Upper Cretaceous Nanaimo Group, British Columbia and Washington State: Canadian Journal of Earth Science, v. 15, p. 405-423.

Mustoe, G.E. Dillhoff, R.M., and Dillhoff, T.A., 2007, Geology and paleontology of the early Tertiary Chuckanut Formation: in Stelling, P., and Tucker, D.S., eds., Floods, Faults, and Fire: Geological Field Trips in Washington State and Southwest British Columbia: Geological Society of America, Field Guide 9.

Savage, N.M., 1984, Late Triassic (Karnian) conodonts from Eagle Cove, southern San Juan Island, Washington: Journal of Paleontology, vol. 58, p. 1535-1537.

Savage, N.M., 1983, Late Triassic (Karnian) conodonts from northern San Juan Island, Washington: Journal of Paleontology, vol. 57, p. 804-808.

Shaw, John Damon, 1972, Late Pleistocene paleontology of Orcas, Shaw, Lopez and San Juan islands of the San Juan archipelago: Master of Science thesis, University of Washington, 60 p.

Shelley, J.R., 1971, The geology of the western San Juan Islands, Washington State: Master of Science thesis, University of Washington, 64 p.

Vance, J. A., 1975, Bedrock geology of San Juan County: *in* Russell, R. H., editor, Geology and water resources of the San Juan Islands, Washington, Water-Supply Bulletin , Washington State Department of Ecology, vol. 46, p. 3-19.

Vance, J. A., 1977, The stratigraphy and structure of Orcas Island, San Juan Islands: in Brown, E. H. and Ellis, R. C. editors, Geological Society of America, Geological excursions in the Pacific Northwest, p. 170-203.

Ward, P.D., 1978, Revisions to the stratigraphy and biochronology of the Upper Cretaceous Nanaimo Group British Columbia and Washington State, Canadian Journal of Earth Sciences, vol. 15, p. 405–423.

Ward, P.W., 1973, Stratigraphy of Upper Cretaceous rocks on Orcas, Waldron and Sucia Islands: Master of Science Research Paper, University of Washington, 44 p.

Ward, P.W., 1976, Upper Cretaceous Ammonites (Santonian-Campanian) from Orcas Island, Washington: Journal of Paleontology, vol. 50, No. 3, p. 454-461.

Weymouth, A.A., 1928, The cretaceous stratigraphy and paleontology of Sucia Island, San Juan Group, Washington: MS thesis, University of Washington, 56 p. .

Whetten, J.T., 1975, The geology of the southeastern San Juan Islands: in Russell, R.H., ed., Geology and water resources of the San Juan Islands, Washington State Department of Ecology, Water-Supply Bulletin, p. 41-57.

Whetten, J.T., Jones, D.L., Cowan, D.S., and Zartman, R.E., 1978, Ages of Mesozoic terranes in the San Juan Islands, Washington: Mesozoic paleogeography of the western United States: D.G. Howell and K.A. McDougal, eds., Proceedings of the Pacific Coast Paleogeography Symposium, Society of Economic Paleontologists and Mineralogists Pacific Section, vol. 2, p. 117–132.

Whetten, J.T., Zartman, R.E., Blakely, R.J., and Jones, 1980, Allochthonous Jurassic ophiolite in Northwest Washington: Geologic Society of America Bulletin, vol. 91, p. 359–368.

INDEX

CPSIA information can be obtained at www.ICGtesting.com
Printed in the USA
BVOW10s0010050515

398896BV00003B/6/P